GENDER, PLANNING AND HUMAN RIGHTS

Gender, Planning and Human Rights explores the geographies and spatialities of human rights with particular emphasis on the connections between gender and human rights in planning and development. Challenging the traditional treatment of human rights cast in purely legal frameworks, the authors argue that, in order to promote the notion of human rights, its geographies and spatialities must be investigated and be made explicit. The book contains a wealth of case studies which examine the significance of these components in various countries with multicultured societies and identify ways to integrate human rights issues in planning, development and policy-making.

The book begins by highlighting the relationships between gender, planning and human rights through a literature review on each of the themes and by making methodological connections. The second section highlights notions of power and control as dominant factors in planning, analysing the relationships between gender, planning and human rights using case studies from the UK, Israel, Canada and Singapore. The final section discusses gendered human rights in development and policy-making processes through case studies in the USA, Peru, European Union, Australia and the Czech Republic.

Presenting challenging insights from the contributors, together with their wide-ranging case study material, this book offers a strong foundation for a new agenda for planning and development which is sensitive to gendered human rights.

Tovi Fenster is a lecturer in Geography and Planning at Tel Aviv University, Israel.

INTERNATIONAL STUDIES OF WOMEN AND PLACE

Edited by Janet Momsen, *University of California at Davis*
and Janice Monk, *University of Arizona*

The Routledge series of *International Studies of Women and Place* describes the diversity and complexity of women's experience around the world, working across different geographies to explore the processes which underlie the construction of gender and the lifeworlds of women.

GENDER, PLANNING AND HUMAN RIGHTS

Edited by Tovi Fenster

London and New York

First published 1999
by Routledge
11 New Fetter Lane, London EC4P 4EE

Simultaneously published in the USA and Canada
by Routledge
29 West 35th Street, New York, NY 10001

Typeset in Baskerville by
RefineCatch Limited, Bungay, Suffolk
Printed and bound in Great Britain by
Creative Print and Design (Wales), Ebbw Vale

British Library Cataloguing in Publication Data
A catalogue record for this book is available from the British Library

Library of Congress Cataloging in Publication Data
Gender, planning, and human rights/[edited by Tovi Fenster].
    p.   cm. – (International studies of women and place)
(alk. paper)
    1. Women in development.   2. Women–Social conditions.   3. Women–
Economic conditions.   4. Women's rights.   5. Human rights.
6. Social planning.   7. Economic development–Social aspects.
I. Fenster, Tovi.   II. Series.
HQ1240.G4545   1999   98–28666
305.42–dc21

              ISBN 0–415–15495–2 (hbk)
              ISBN 0–415–15494–4 (pbk)

CONTENTS

CONTRIBUTORS

Gillian Davidson is a geographer who has conducted research on gender, poverty and migration in South-east Asia. Until recently she researched at the Centre for Applied Population Research in Dundee, Scotland, and is currently conducting socio-legal research for the Law Society (England).

Tovi Fenster is a Lecturer at the Department of Geography, Tel Aviv University, Israel. She has published articles and book chapters on ethnicity, citizenship and gender in planning and development. Currently she is writing a book entitled *Gender, Space, Culture in Planning and Development* (Longman). She is a member of the board of directors of the Association for Civil Rights in Israel.

Eleonore Kofman is Professor of Human Geography at Nottingham Trent University, England. She has published extensively on citizenship, welfare and the development of Fortress Europe. Currently she is co-authoring *Gender and International Migration in Europe* (Routledge) and writing a book titled *Political Geography* (Routledge).

Maureen Hays-Mitchell is an Assistant Professor of Geography and Coordinator of the Latin American Studies Program at Colgate University, in the USA. Her research interests include gendered dimensions of economic restructuring in Latin America, with particular emphasis on the urban informal and micro-enterprise sectors. She has conducted field research in Peru, Mexico, Chile and Spain.

Jo Little is a Senior Lecturer in Geography at the University of Exeter, England. Previously she taught town and country planning at the University of the West of England. She is the author of *Gender, Planning and Policy Process* (Pergamon) and co-editor of *Women in Cities* (Macmillan).

Beth Moore Milroy has a doctorate in city and regional planning. Her teaching and research fall mainly in the areas of feminist planning theory, participatory planning practices, community work, and the street as civic realm. She is Director of the School of Urban and Regional Planning at Ryerson Polytechnic University in Toronto, Canada.

Ann M. Oberhauser is an Associate Professor of Geography in the Department of Geology and Geography at West Virginia University in the USA. Her research interests include gender and economic restructuring, particularly rural women's home-based economic strategies and flexible networks in Appalachia and southern Africa.

Deborah Bird Rose is a Senior Fellow of the North Australia Research Unit, Institute of Advanced Studies of the Australian National University. Her current research focuses on indigenous and settler cultural and ecological encounters. She is the author of *Nourishing Terrains: Australian Aboriginal Views of Landscape and Wilderness*; *Dingo Makes Us Human* (winner of the 1992/3 Stanner Prize); and *Hidden Histories* (winner of the 1991 Jessie Litchfield Award).

Jiřina Šiklová is a Professor of Sociology and the head of the Department of Social Work, Charles University, Prague. She has written in depth on social problems and women's problems in Communist and post-Communist regimes. She is currently a consultant for government offices such as the Czech Academy of Science, the Ministry of Labour and Social Affairs and the Ministry of Culture.

Marcia Wallace has a Ph.D. in Urban Planning from the University of Waterloo, Ontario, Canada. Her specialization is on the challenges of multicultural diversity and immigration for urban and suburban planning. Other research interests include local citizenship, public participation and community development.

ACKNOWLEDGEMENTS

I wish to express my deep gratitude to Janice Monk and Janet Momsen for their support, editorial advice and assistance during the process of editing the book, to the contributors for their patience and to the publishers Tristan Palmer and Sarah Lloyd for their encouragement.

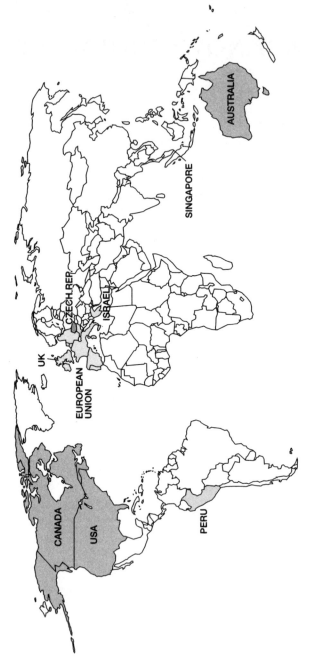

The countries and regions referred to in this book

Part I

INTRODUCTION

1

GENDER AND HUMAN RIGHTS: IMPLICATIONS FOR PLANNING AND DEVELOPMENT

Tovi Fenster

This book is about the geographies and spatialities of human rights. It aspires to challenge the traditional treatments of human rights that are cast exclusively in legal frameworks by arguing that, in order to promote the notion of human rights, its geographies and spatialities must be investigated and become more explicit. The rationale behind this argument is that space is more than relevant for understanding human rights violations. Some of the most brutal and cruel cases of human rights abuse are connected to the lack of freedom to move in space by imprisonment at home, whether it is enforced physically or psychologically through fear and terrorism or imposed by rules and the cultural meanings of spaces. One of the major concerns of the human rights discourse is the practice of locking up women or prohibiting them from moving freely in their environment in order to ensure their faithfulness and modesty, while their husbands are free to move around. The right to work and the right to political participation are also abused because of lack of freedom to move in space (Nussbaum 1995). Fear of violence makes women avoid certain spaces. Should women go to a bar for a drink, as men do, the response may be sexual terrorism. Men's control over public space in the evening makes even western women fear male violence, deterring most of them from being independent despite their professional success. Women's inhibited use and occupation of public space is seen as a spatial expression of patriarchy (Bunch 1995, Valentine 1989) and as a violation of human rights. This situation shows that space is never neutral; instead, it affects and is affected by social and power relations in society.

That is why, until now, voices criticising the legal exclusivity in human rights discourse have argued that male dominance such as this makes freedom of movement impossible and encroaches upon socio-economic rights. Hence, other approaches, apart from litigation, are needed to promote human rights (Gomez 1995). This is exactly what this book suggests. It provides another, perhaps new, channel for promoting and implementing human rights in planning and development – via understanding of their spatialities.

The discourse about planning, especially for multicultured societies, has

3

become very complex and politically charged in many parts of the world since the mid-1970s. For this very reason, this book highlights human rights issues in relation to planning and development. Spatial planning is never a neutral process. It reflects social and power relations within a society as well as affecting them and, to a large extent, spatial relations actually represent and sometimes also reproduce social relations (Moore 1996). It is via planning schemes and development projects that states' policies are carried out, with or without attention to human rights issues, affecting every aspect of daily life. Therefore it is important to focus on the incorporation of space into human rights discourse. The need for such attention is most acute for members of ethnic minority groups in multicultured societies as well as for immigrants.

The geographies and spatialities of the human rights discourse are tackled here in several ways; first, the chapters represent different geographical settings. Each looks at the human rights discourse in a different country, culture, political and social background. Second, the case studies represent human rights issues in a range of political settings such as liberal western democracies (the USA, UK, Australia, Canada, Europe and Israel), as well as in a post-Communist regime (the Czech Republic), and in semi-totalitarian regimes (Singapore and Peru). What comes out of these experiences is that in spite of the different geographical and political settings, abuses of women's human rights appear to be similar. Third, the book challenges the extent to which geographical divisions such as developed/developing, private/public and other dichotomies are relevant to the analysis of human rights issues. As the case studies show, these dichotomies disappear when looking at how women's human rights are abused. Fourth, the chapters highlight the different scales in which human rights issues can be analysed, especially with regard to planning and development: the body, the home or the private space, the street, the neighbourhood, public spaces, regions within countries and between countries, and national and international levels. Taking these various scales into account not only sheds new light on identifying obstacles in promoting human rights issues at each scale but also reveals the links between abuses of human rights across the different scales. Pointing out the commonalties and differences between the case studies offers new insights enabling us to formulate more effective ways to promote human rights issues. Some of these understandings will be presented in the conclusions of this book.

Each of the fields addressed in the book – of gender, planning and human rights – can be viewed as a topic in itself in terms of its political agenda and the practical implications for individual life; indeed a growing body of literature exists on each of these themes. However, the intricate links between the three, that is, how gender and human rights should be expressed in planning and development, are less investigated dimensions of the subject. This introductory chapter aims at highlighting some of the issues closely related to each of the three themes and their interrelationships. Since there is some confusion in the notion of human rights, I will begin by presenting some of the definitions used in the literature.

DEFINITIONS OF HUMAN RIGHTS

Human rights are defined at different scales, and in relation to different spaces and spatialities (socially constructed spaces), with the human body being the basic 'territory' in the discourse. Physical violation, killing, imprisonment, torture and rape are only a few examples of human rights violations of the 'body'. Nevertheless, acknowledgement of bodily rights was not explicit in the United Nations Universal Declaration of Human Rights of 1948, the first UN declaration on human rights. This declaration 'recognises the inherent dignity and equal and inalienable rights of all members of the human family as the foundation of freedom, justice and peace in the world'. Although it does reaffirm 'the faith in fundamental human rights, in the dignity and worth of the human person and in the equal rights of men and women' (in Wetzel 1993), it is still rather vague and general about the rights of the human body. Probably, as in many other UN declarations, vagueness is sometimes necessary to make these declarations accepted internationally. Nevertheless, this document became a means for the UN apparatus to ask countries questions about human rights issues that were previously considered internal affairs. From then on a large number of documents, declarations, charters and covenants has been produced. As a result, an impressive body of international law looks at human rights in different sectors of life: economic, housing, political, cultural, social and spatial as well as at different spatial scales: the body, home, the street, the neighbourhood, public spaces, regions and countries.

When the human body is taken as the basic scale in the human rights discourse gender becomes a basic factor. While men suffer from similar abuses of their human rights, women are usually more vulnerable. Their dignity, freedom and equality is more readily violated than those of men partly because traditional human rights formulations are based on a 'normative' male model and applied to women as an afterthought, if at all (Charlesworth 1995). This male bias raises a concern about the lack of representation of feminist perspectives in the formulation of international human rights standards (An-Na'Im 1995).

The concept of human rights as women's rights in relation to their bodies refers mainly to the right to possess one's own body, including reproductive rights, freedom of sexuality, freedom of marriage, laws that criminalise rape in marriage, and laws that criminalise forced prostitution. These human rights issues are, sadly enough, the arenas of women more than of men. As the next section reveals, women's human rights are constantly abused, and women are always more vulnerable. In such a situation, a gendered analysis of human rights issues is indeed justified.

With regard to the gender-neutral (or blinded) human rights definition, some argue that what is included in the definition of human rights is what encompasses basic human needs, i.e. the right to work, to adequate food, to education, to shelter, and to health (Wetzel 1993). Broader definitions invoke issues of empowerment (Cook 1995), including participation in the development and planning of one's environment. In this context, participation may as well be viewed as part of citizens' rights and as part of people's full membership in a community (Lister 1995, Marsall 1950). Participation is a contested term, as its validity relies substantially on the political context of its implementation. It is indeed crucial to involve people in the planning process, but it is not always enough to ensure the incorporation of human rights issues in planning and development. To guarantee that, participation must be about empowerment, that is, there must be a real commitment of policy- and decision-makers to empowerment (Howard 1995).

Another question under debate is whether 'human rights' is a different concept from 'citizenship'. Wetzel (1993) argues against this position, proposing the compatibility of human rights with state citizenship. In his view, the human rights system is built upon the notion of an organised state, a society that is obliged under the human rights law to provide its citizens with freedom from violation of each individual's rights. From this perspective, the state is required to create conditions under which its citizens can meet their needs expressed in the human rights system. In contrast, Lister (1995) suggests that the notion of human rights is more global than that of citizenship rights because it relates to both citizens and non-citizens: immigrants, refugees, asylum seekers, for whom human rights are even more crucial than for citizens. Without this distinction, she argues, human rights become an exclusive right of nation-states' citizens whereas non-citizens find themselves excluded.

ARE HUMAN RIGHTS GENDERED?

As already mentioned, human rights are tragically gendered on all scales of analysis. Women and their dependent children make up most of the world's refugees and displaced people (Amnesty International 1995). Human rights are also gendered, since women represent a large majority of the poor in every country. To acknowledge economic well-being as a human rights issue would certainly help to eradicate poverty. The exclusion of women from positions of public power, especially at the national scale, is another aspect of gendered human rights. This exclusion prevents them from being involved in decision-making in shaping laws and institutions affecting women's (and others') lives.

The following somewhat historical background of the development awareness of the gender dimension of human rights shows up the dominance of legal aspects in this discourse. Officially speaking, the gendered nature of human rights has been validated by special UN events held since the beginning of the 1970s. The proclamation of the year 1975 as the International Women's Year marked the start of several UN conferences on women's issues. The first, held in 1975 in Mexico City, produced the Declaration of Mexico, 1975. It was a revolutionary conference that recognised for the first time the oppression of women everywhere, linking this situation of oppression with inequality, with underdevelopment as well as with an unjust world economic system. In addition, the Mexico declaration mandated the elimination of such violence against women as rape, incest, forced prostitution, physical assault, mental cruelty and coercive and commercial marital transactions. Though it has been adopted by majority vote, the document was largely ignored by most nations. Its importance may lie in the fact that it helped women to realise that collective action was the key to their power and effectiveness; the ensuing links of solidarity began to forge a chain of women throughout the world.

The UN in Copenhagen, Denmark endorsed the subsequent Convention on the Elimination of All Forms of Discrimination against Women in 1980. It contributed to the recognition that discrimination against women is a social problem that requires urgent solution. This Convention suggested a set of corrective measures which must be implemented in each state, such as the passage of measures to ensure women's equality in the legal system, guaranteeing women the right to go before tribunals and other public institutions when faced with discrimination, and so on. The problem remains, however, that no mechanisms exist to enforce the Convention within states, so that these conventions have changed the lives of women little or not at all. It is now well recognised that the implementation of these conventions depends largely on both global movements of feminist international power and local movements of women themselves, as well as on their awareness of the existence of these conventions.

The final UN Decade of Women conference in Nairobi in 1985 emphasised the global power of women. Fifteen thousand women from over 150 nations stood united despite their national, ethnic, racial, geographical, cultural, economic and age diversity. The Nairobi Forward-looking Strategies for the Advancement of Women (1985) reflected the commonalties of experience and needs, pointing to the fact that universal oppression and inequality are grounded in the patriarchal systems that ensure the continuance of female subservience and secondary status. This document also acknowledged the fact that women do two-thirds of the world's work, yet two-thirds of the world's women live in poverty, their work is unpaid, underpaid and invisible; women are peace-makers, yet they have no voice in arbitration; there is a universal sexual exploitation of girls and women, too often resulting in sexual domination and abuse throughout their lives; women provide more health care (both physical and emotional) than all the world's health services combined; women are the chief educators of the family, yet they outnumber men among the world's illiterates at a ratio of three to two. Even when educated they are not allowed to lead. This scenario shifts culture by culture but the story line remains the same (Schuler 1993). In practice, however, gains that achieved even *de jure* (legislative) equality were not matched by *de facto* reality.

The International Women's Rights Action Watch, set up in 1986 is one of the mechanisms established to monitor progress towards the achievement of human rights for women. The presence of women from every corner of the Earth at the UN Human Rights Conference in Vienna in 1993 sent a further clear sign that women had entered the formal human rights arena and demanded that their human rights be met.

The Beijing Platform for Action of 1995 called for the government's active participation to end discrimination. It demanded that governments promote, increase, provide or ensure the availability of and access to, the health care they need, the education that is required for literacy, and move women out of poverty, end violence against them and eliminate sexual harassment. Improvements in women's lives were seen in Beijing as benefiting society at large, that is, improve women's lot and society as a whole will benefit – a new language, a new thought. In this conference, the issue of government's 'right' to interpret human rights according to their own philosophy or circumstances has emerged with the concern that this could be a regressive step with regard to women's human rights issues.

In spite of the international nature of the concept, gendered human rights may have different meanings for women from different parts of the world and of different political backgrounds. For women in post-Communist countries of Eastern Europe, for example, human rights discourse may be more promising than the discourse of western feminism, because women's rights, like political participation, is overly identified with former Communist regimes; the human rights discourse can be used to bring attention to transitional violence against women (Eisenstein 1996). This point of view is expressed in Jiřina Šiklová's chapter on the Czech women in the post-Communist environment. Cultural and political differences highlight a very basic question about the expressions of the human rights discourse at different scales and in different spaces such as the contested 'private/public' divide. What is 'public', what is 'private' and who defines the boundaries between them is a geographical as well as political and cultural issue.

GENDER, PLANNING AND HUMAN RIGHTS

Planning and human rights

Planning is defined here as a set of rational actions, which aim to organise the use of space according to principles and goals determined in advance, usually by those in power. As such, planning is a very powerful action since it dictates the future of individuals and groups by shaping economic, social, cultural and physical spaces, which usually meet only the needs of the powerful actors in society. This interpretation follows the Marxist argument that spatial planning is one of the expressions of power relations between the different social groups within society, since the goals for planning are usually formulated by the dominant group controlling the resources (Paris 1982). One of the powerful tools for control of development and use of land is the zoning system. Huxley (1994) argues that zoning is used to create homogeneous areas that exclude 'the other' as part of the state's involvement in defining and control of citizens/subjects. The dominant group, that is, politicians and planners, decides upon principles of zoning, because it possesses the knowledge and this knowledge provides it with the power, which, in turn, produces knowledge (Foucault 1977). This interpretation reflects Foucault's thinking about relationships between discourse, power, control and manipulation of knowledge and ultimately of society (Moore 1996). From this perspective, planning can be perceived as a spatial mirror of social relations, that is, those who dominate socially and culturally also dictate principles of planning. Foucault views text as a means of concealment and of deliberate control. Planning procedures, documentation and the actual plans themselves can be seen in this light as texts consisting of complicated language, known only to professional planners, where the hidden agendas underlying their creation form a 'knowledge' shared only by exclusive people and not among the public mostly affected by these plans. This knowledge creates the power to control space and exclude or include human rights principles. In this respect Huxley (1994) finds similarities between planning regulations such as zoning and Foucault's views on the discourse of imprisonment and social control in Enlightenment logic, a discourse that also generates 'the other'.

The connection between planning procedures and human rights is, then, in the *way* plans are being made, that is, the extent to which their methodologies dictate the prospect of both promoting and internalising human rights concerns, or denying or ignoring them. The planning process or procedure clearly determines or reflects the type of relationships that exist between the state and its citizens, including their human rights.

Two approaches have been identified for analysing planning theories and procedures with regard to social relations and space, the *procedural* and the *progressive* (Alterman 1994). The former represents a modernist outlook on society, a formal, top-down process that ignores the 'others' – the marginalised – and pays less attention to social relations, citizen and human rights and to their expressions in space. Procedural planning theories view society as a homogeneous entity, rather than looking more deeply into its social and cultural structures. They rely on functionalist, modernist thinking to achieve planning goals and employ a rational comprehensive methodology.[1] In many ways this approach views planning procedure as an end in itself. In contrast, progressive planning represents a more open view of social structures and relations, and connects them directly to the use of space. It is not merely a formal top-down exercise, but instead involves an ongoing dialogue between institutions and beneficiary groups. This type of planning serves as a means for achieving social and political goals, building beneficiary capacity, beneficiary empowerment and creating equal access to resources. Such an approach includes advocacy planning, negotiated planning, critical planning and radical planning (Alterman 1994). Progressive planning derives from a postmodernist view of society, which challenges the 'grand theory' of modernism, emphasising the particular and the local. The two planning traditions reflect different perspectives on citizenship and human rights issues. A more rigid, controlled, centralised and patriarchal relationship between state and citizen is evident in the procedural planning tradition, while the emphasis on human rights, citizen rights, and participation is a necessary base for the progressive planning tradition.

Another classification of planning traditions identifies three main groups

9

(Safier 1990; also in Moser 1993). First, are the 'classical' traditions concerned with physical and spatial problems of city and regional growth with the 'blueprint' approach as its methodology.[2] This tradition and methodology clearly emphasises the knowledge–power relationships in planning because it assumes that state bureaucrats can be and are the ultimate authority for translating the information about and understanding of economic, social and environmental needs into physical and spatial forms. Second are the 'applied' planning traditions, which shifted their concern from spatial and physical domains to economic and social growth at both project and corporate levels. These applied traditions are identified with a rational comprehensive methodology. This methodology, still widely used, has been criticised for focusing on the means of planning, rather than the end products. The rational comprehensive method also emphasises the power relations within the planning process as well as the fact that each of the logical stages in this approach is dominated by decision-makers and planners while voices of the 'beneficiaries', the 'public' or the 'people' are unheard. Advocacy planning, negotiated planning and so forth, are alternative methodologies to the rational comprehensive. These methodologies allow the 'voices from below' of 'the other' to be heard and the 'power game' between individual or group and the elite takes a more dominant and explicit role. Finally, in this categorisation of planning traditions, it is the 'transformative' tradition that requires much more transformative procedures than the other pervasive methodologies (Moser 1993). Transformative tradition includes development, cultural, environmental and gender planning, and is obviously more sensitive to human rights issues then the first two.

Gender has emerged as a powerful and transformative theme in urban planning in recent years for numerous reasons, such as the growing feminist movement, the entry of women into the paid labour force, the increasing number of women planners and professors, and the parallels between the civil rights movement and the feminist movement. Feminist perspectives address many issues of urban life, challenging the homogenisation of spaces that is a result of planning procedures and regulations: the safety of women in cities, structural discrimination against women in economic development, the transportation needs of women beyond the traditional 'journey to work' and the impacts of traditional suburban housing on antiquated nuclear family structures. Since planning is inherently political, planning theory and practice need to attend to the debate about women's human rights. Marcia Ritzdorf (1996), a US planner, has found that women have been generally interested in expanding the range, intensity and modes of action in planning. They are interested in holistic approaches to problems and comparative problem-solving and see issues impacting their bodies, their families and their neighbourhoods as both political and personal.

What is the significance of feminist contributions to planning theory? Friedman (1996) recalls the radical planning tradition to which feminist thought is the most relevant. This tradition, which is part of the progressive planning approach, looks at planning as coming from the civil society and being concerned with the emancipatory movement. It is very much on the political agenda and has become a widely shared ideal. It is a planning with and for the civil society, especially for those sectors that have been silent and submerged. Such feminist thought comprises an attack on male domination. Within the urban realm, feminists have contended that men design cities to serve male needs (Fainstein 1996). Even when women occupy positions as planners, they do so on terms dictated by men and lack the power to reorient the planning project so as to accommodate the interests of women. In her chapter in this book, Jo Little emphasises how gender inequality in the built environment is caused by the reluctance of the planning apparatus to develop women's initiatives in the UK, clearly showing how women's human rights are abused because of their limited ability to exercise their power.

All categorisations of planning reflect different perspectives on knowledge and power relations as well as on social relations. These different perspectives clearly affect the ways in which human rights issues become an integral part of planning and development. Caroline Moser (1993) suggests how to adopt a new gender planning methodology, which is both political and technical in its nature, based on 'negotiation' as a means of meeting women's human rights in planning and development. Such a planning methodology can become an efficient channel for promoting and implementing human rights for both men and women.

Development and the human rights discourse

Just as approaches to planning are debated and have diverse implications for achieving human rights goals, so too do concepts of development (Young 1993). While some critics see the hidden agenda of 'development' as nothing less than the westernization of the world (Sachs, in Gasper 1996), it is more commonly perceived in more neutral terms as combination of growth, self-reliance and equitable distribution of resources to 'all citizens'. Going further, neo-Kantianism would conceptualise development as extending ability of individuals to set and pursue ends freely chosen by them in order to meet their human needs and enjoy their human rights. Within the discourse of planning and development, concepts of human rights usually incorporate rights to an equal voice, literacy, information, political participation and public accountability as well as the right to equal access to development components, such as health, food, water, education, housing, employment, land and credit (Howard 1995). Human rights in planning refer also to the right to equal opportunity in initiation in the public sector of planning and development.

How the two general concerns of development expressed above, economic betterment and greater equity, should come together has not been resolved yet in the development discourse. For some, economic growth is the primary goal that comes before and precludes considerations of human rights. For feminists, it is important to challenge this position, since considerable evidence exists that it is women who frequently suffer severe economic and social dislocation as a result of 'development'. Even the 'women in development' approach is criticised for its depoliticised stance focusing on economic rather than power issues (Howard

11

1995). Concepts of development are also attacked on grounds that they proffer a universal concept of social improvement, which does not adequately acknowledge cultural norms of development. This is especially expressed in the UN Declaration on the Right to Development. This Declaration announces development as the right of 'every human person and all peoples . . . to participate in, contribute to and enjoy economic, social, cultural and political development in which all human rights and fundamental freedoms can be fully realised'. The right to justice in planning means that the goals and targets of planning will be determined not only by the elite professionals but also by the people who presently and in the future are affected by these planning procedures. This is one example of the ways in which human rights in planning and development can be achieved. The participation of people is a necessary determinant of human rights in development. Participation is also part of citizen rights, the right to influence one's life and the right to knowledge and involvement in determining the future of a spatial activity. Nevertheless, although participation is crucial, it is not sufficient to ensure women's human rights in planning and development. As Rhoda Howard (1995) says, women's integration into development projects does not necessarily improve their overall status.

THE SPATIALITY OF THE HUMAN RIGHTS DISCOURSE

Human rights and the private/public discourse

The distinction between private and public spaces provides insights for the better understanding of human rights violations. In the same way it shows how tricky it is to apply western theories to other social and legal systems (Cook 1995). The large number of definitions of and perspectives on the two dimensions indicates how much they can be confused. First, there is the culturally oriented approach which claims that boundaries between the private and the public vary according to cultural, political and geographical settings. Some writers refer to the public sphere as identical to the political sphere, while the private sphere relates mainly to the family domain where women are primarily located (Cook 1995; Yuval-Davis 1997). In any case, its contents and boundaries may be different in different cultures and in different contexts precisely because private/public divides represent projected cultural meanings and perceptions of space. Charlesworth (1995) suggests that 'what is public in one society may well be private in another'. What is probably universal about the private/public divide is that it represents, reproduces and reinforces patriarchal power relations of gender, race and class (Sullivan 1995).

Second, is the argument that the distinction between 'private' and 'public' comes from the western liberal thought. This distinction has its roots in the transformation of the hegemonic power relations in the society from a patriarchy, in which the father rules over both other men and women, to a fraternity, in which the men have the right to rule over their women in the private domestic sphere, but agree to be socially equal in the public, political sphere (Pateman 1988, 1989). This means that actually nothing has changed.

Third, there is the feminist point of view, which relates 'private' spaces within the human rights discourse first and foremost to the rights of women in relation to their private 'territories' – their bodies. These include the right to prevent violence against their bodies such as rape, sexual harassment, forced prostitution; the fight against discriminatory family and marriage laws and against any restrictions on women's rights and the condemnation of any persecution on the basis of sexual orientation. To fight these abuses, feminist activists are arguing that the dichotomy of public/private is false, and is invoked largely to justify female subordination and to exclude the abuses of human rights at home from public security. Feminists further argue that what occurs in private spheres shapes women's abilities to participate fully in public activities and that women who are subordinated in private cannot be full and equal participants in public matters (Bunch 1995). They also criticise the gendered aspect of the private/public dichotomy because it rests on basic assumptions about women's particular role in biological reproduction and their subordinate role within the economy (Eisenstein 1996). To add another dimension to this discourse, Binion (1995) suggests that the state is not the only powerful actor to want to limit the reach of human rights to the 'public' domain, since the pressure to do so might also come from the private domain through religious institutions and corporations. These bodies have much to gain from preserving their autonomy in both private and public domains and their dominance increases the restrictions on women's spatial mobility. As a result, women in countries with Islamic fundamentalism on the rise are forced to return home or to wear a veil in public because the notion of public and private has changed so that it limits their freedom to move in the public domain. Feminists argue that this dichotomous formulation leads to many of the violations of women's human rights in the private sphere because it is by definition excluded from the work of international and non-governmental organisations.

The notion of public (as well as private) is not only a gendered matter but also a race and class matter (Sullivan 1995). In many cases, the notion of public interest is being replaced by the notion of 'the wealthier upper class public interest' (Eisenstein 1996). Public spaces are renegotiated for the interests of only some members of the 'public', and power relations define the boundaries between private and public. A clear example of such contradictions is the reduced investment by governments in public services such as playgrounds or parks in poor and minority neighbourhoods while tax breaks and other public mechanisms facilitate gentrification and up-scale developments for the affluent 'public'. Thus, boundaries between the private and public become matters of class, cultural, political and social settings. They may change among different classes, races, and geographical regions within a country, between urban and rural environments (Sullivan 1995) and between densely populated or dispersed areas.

The fact that participation in economic, social and political activities is perceived in many societies as public limits women's access to and control of

13

resources. The primary identification of women with the private sphere of the family and home contributes to their secondary status in the public, the very realm where their future is debated and even decided upon without their being involved in political decision-making processes (Rao 1995). Such lack of access and involvement affects women's roles and participation in development and planning as much as in other spheres of life. Within the planning apparatus, the geographical separation between private and public perpetuates traditional perceptions of gender activities. Residential zoning exacerbates the division of home from work, as well as the notion of social services as 'soft' in contrast to infrastructure and buildings as 'hard' investments.

Perhaps the most explicit evidence of the 'elasticity' and contradiction between these definitions is the fact that new practices and new scales of organisation within global capitalism are currently redefining the public/private divide in both First- and Third-World countries. Mega-corporations are replacing the power of governments as economies are being built on a global scale. In this context, private/public divides are no longer clear-cut (Eisenstein 1996) because services, spaces and activities such as health, childcare and care of the elderly that used to be public in industrialised welfare states are now privatised. In developing countries, the opposite is occurring; what was the duty of the family is now in the hands of the state (child and elderly care in Third-World countries) while global corporations have taken over employment that was formerly locally organised. The processes of transitional economic intrusion into Third-World countries means in many cases, further exploitation of women as poverty-wage workers, particularly in factories. Another example of the 'invasion' of the public into the private is the increasing practice of homeworking, which further exploits women's labour. This new situation calls for the necessity of a new approach regarding this dichotomy, and Yuval-Davis's (1997) suggestion that three spheres should be differentiated – the state, the civil society and the domain of the family and kin – might be a starting point for a new vision.

GENDER, HUMAN RIGHTS AND CULTURE

Cultures operate within both social and spatial contexts within the time dimension (Massey 1994). That is, different positioning, both socially and geographically, affects the ways cultures are articulated and used, both inside and outside collectivities (Yuval-Davis 1997). This situation poses a dilemma: in a world of so many cultural differences is it possible to discuss human rights issues internationally?

An overly simple notion of the relationship between culture and human rights in the world of differences could result in a dichotomous formulation, with universalists falling on one side and relativists on the other (Donnelly, Manglapus, in Rao 1995). While universalists argue that human rights are universal, that is, considered by the international law as the rights held equally by every individual, relativists argue that efforts to 'universalise' human rights can be viewed as an act of imperialism or colonialism. As Radhika Coomaraswamy (1993) put it, the discourse between the two approaches is a battle between those who are strug-gling for democracy, pluralism and ethnic tolerance and those who are strongly nationalistic or believe in return to religious fundamentalism where the com-munity of interests is laid down at birth and by tribe and not on the basis of what is fair, just and equal.

In reality, the universality of human rights is usually undermined by govern-ments who argue that human rights must be subject to the interests of national security, economic strategies and local traditions. When it comes to women's human rights, many governments take a particularly restrictive view with regard to women's equality and freedom to move, or their rights to have free access to resources, because women represent the reproduction of the family, the code of honour and modesty of a society and a nation. Thus, women's human rights abuses are often justified in the name of cultural and moral codes. Through their clothing and demeanour, women and girls become visible and vulnerable embodiments of cultural symbols and codes (Rao 1995). One such example is the restriction of women's freedom to move in public because of cultural codes. This example brings us back to the very basic questions of who defines the spatial, social and political boundaries between the 'private' and the 'public' and how dual legitimacies of political and cultural considerations should be adjudicated. How should they be expressed in defining women's access to space and in taking account of their needs in planning projects and development schemes? In some Muslim societies, what is defined as 'public' space by men and for men is labelled as 'spaces of immodesty' for women, and the freedom of women to move in the 'public' is restricted by cultural norms. In western countries, codes of safety and fear turn areas into 'forbidden' spaces for women. As previously mentioned, this situation prevents women from becoming involved in activities outside their homes, perpetuating their subordination and lack of involvement in public affairs and access to resources. For human rights issues to be addressed effectively, they have to become a respected part of a culture and the traditions of a given society (Coomaraswamy 1995). The main concern should therefore be how to legitimise international human rights within particular cultures and traditions. This means that some of the basic cultural settings need to be transformed so that they reflect human rights values (An-Na'Im 1995). In this book we suggest that one way of introducing human rights into cultures is via planning and development, because planning can either adopt or change the cultural meanings and perceptions of space. It is actually the universality versus relativity dilemma that planners face.

For planners, this cultural relativity of human rights present challenges. Should traditional norms and cultural habits causing women's subordination be regarded as 'human rights' in order to be taken into consideration in planning schemes and, if so, to what extent should planning schemes play a role, either in changing these norms or perpetuating them? One way of tackling this discourse is to distinguish between human *rights*, which are universal and determined by international law, and human *needs* which are local in nature and vary from one culture to another. Coomaraswamy (1993) provides another answer for this dilemma. She acknowledges the diversity, but only within the context of a firm foundation of universally acknowledged human rights and women's rights. Perhaps a more just way to decide whether the cultural and political priorities of a particular government or a particular patriarchal community should take precedence over the collective will of the international community, is to ask what women want and think. As Amnesty International (1995) argues, governments do not have the authority to define what constitutes a fundamental human right or who may enjoy that right.

THE STRUCTURE OF THE BOOK

The aim of this book is to explore the geographies and spatialities of human rights with particular emphasis on the connections between gender and human rights in planning and development. The authors discuss the significance of these components in various countries with multicultured societies, and identify ways to integrate these issues into planning, policy-making and development strategies. They address the implications of planning for both women and men in majority and minority groups, and analyse the role of the state in the process. Their analyses can then form the foundation of a new agenda for planning and development, which is sensitive to gendered human rights.

The book has three parts. Part I highlights the intricate connections between the three leading dimensions of the book by reviewing the existing literature on each of the themes and making the methodological links between these dimensions. Also, a discussion on private/public divides and the notion of culture within the human rights discourse is presented.

The four chapters in Part II analyse the relationships between the three leading dimensions of the book: gender, planning and human rights as they are applied in four different geographical settings, in the UK, Israel, Canada and Singapore. In Chapter 2 Jo Little departs from the traditional focus of 'women and planning' discussions to provide a more critical analysis of the failure of planning to advance women's human rights. She analyses how the lack of women's representation in official planning bodies decreases the chances of meeting women's rights in town planning, a field that has the ability to influence key aspects of everyday life in which women's human rights are frequently abused: aspects such as accessibility to resources, employment and housing. Her analysis attempts to explain the failure to meet women's human rights in planning by referring to the operation of power within the decision-making process and to the wider relationship between planning, the local state and local political activity. She focuses on issues of gender and power in planning by reviewing women's initiatives through formal planning policy and by looking at the formal structures, that exist in local authorities in the UK for addressing women's rights and issues of gender inequality. She places a particular emphasis on the extent to which women's committees within local authorities can stimulate more positive approaches to meeting women's needs in planning policies. She addresses this question through a case study of the changing relationship between planning and women's committees at the Bristol City Council.

In Chapter 3 Tovi Fenster approaches planning as a means of control which abuses minority women's human rights in Israel. The notion of gender and power is expressed in identifying the dilemma of whether to take cultural values into consideration in spatial planning, even if these values abuse women's human rights. The theoretical part of the chapter looks at the relativist versus universalist argument existing in the human rights literature, using them to analyse Israeli state settlement and housing planning projects for two groups: indigenous Muslim Bedouin and immigrant Ethiopian Jews. She focuses in particular on the effects of these plans on minority women and shows that the universalist-modernist planning approach, which was adopted in the two planning projects, created a situation in which Bedouin women were more restricted in their movement in space. This happened because higher housing density in towns increased the chances of breaking cultural codes of women's modesty and therefore more spaces become 'forbidden'. For the Ethiopian women, a lack of consideration on the part of the planners to the need for 'spaces for menstruation and after baby birth' created situations in which 'pure' and 'impure' spaces were mixed, leaving the Ethiopian Jews with hard feelings and breaking their traditions. The two case studies clearly present this ethical dilemma and the negative effects of universalist plans on the situation of women in each group.

In Chapter 4, Marcia Wallace and Beth Moore Milroy address the challenges for planning in multicultural Canadian cities. They challenge the view of diversity as an exception that needs accommodating over and above the 'normal' set of demands and needs covered by planning functions. They suggest that cultural diversity should be recognized as part of the basic make-up of cities and that such recognition means looking at gender, ethnicity and class as intersecting one another. They describe how treating diversity as an exception is currently reinforced by broad societal structures and by planning frameworks and argue that this approach negatively effects meeting human rights in general and women's human rights in particular. Wallace and Moore Milroy emphasise the notion of power in planning and the dilemmas that arise in a country where provinces have ultimate power on land use planning while principles of human rights and multicultualism are established at the national level. They present the case of Ontario, illustrating the squeeze planners find themselves in in practice between the 'generic-leaning' planning legislation on the one hand and the 'diversity-leaning' professional statements of value and human rights decisions on the other hand. They use the example of the Old Order Amish religious community to emphasise this ambivalence.

In Chapter 5 Gillian Davidson challenges official Singapore state planning. She asks how the preoccupation of state planning with both economic growth and multiculturalism impacts the human rights experiences of different groups of women in Singapore. Whilst Singapore is often viewed as an international success story, the human rights of its citizens, especially with respect to equality, inclusion and empowerment, are not necessarily secured. Drawing on two examples from social planning which focus on the definition of what it is to be a 'woman' and a 'mother', the chapter highlights how the construction and inculcation of cultural values, morals and gendered roles as tools for planning impact on the fragile nature of human rights and equality for women. Davidson describes the ways in which the state's construction of cultural and national values includes notions of what is a 'good woman' and what is a 'good mother'. This notion of culture is also strongly connected to that of power in that it is exemplified in the exercise of 'boundary-drawing'. Davidson provides two examples of gender inequality in Singapore's population policies, first in their pro-natalism, which promotes higher fertility and in that way uses women as tools of state planning, and second in policies promoting family life that marginalise single mothers.

Part III focuses on how human rights perspectives and gender issues are expressed in development and policy-making. It looks at the economic rights of Appalachian women in the USA, at the rights of migrant women in the European Union, of indigenous women in Australia, and of women in post-Communist countries such as the Czech Republic. In Chapter 6, Ann Oberhauser analyses human rights and development from a feminist perspective better to understand the links between household relations, gender-based violence, and unequal economic opportunities for women. She argues that gender dynamics in the private domestic sphere can negatively impact on women's participation in the formal labour force and, consequently, reduce their economic status. Her theoretical framework examines gender inequalities in both the private and public spheres of the household and workplace. Oberhauser argues that conventional approaches to development and human rights often ignore the relationship between women's productive and reproductive labour, thus failing to consider how household dynamics contribute to gender-based violence and unequal access to employment. The empirical analysis in the chapter applies these themes to a case study of gender and women's work in the economically depressed region of Appalachia. This discussion draws from socio-economic data on education, employment, and domestic violence and findings from semi-structured interviews with over eighty West Virginia women.

In Chapter 7, Maureen Hays-Mitchell seeks to advance our understanding of gender-based discrimination inherent in the neo-liberal model of development currently pursued by states throughout Latin America and the relevance of this model to the human rights community. To this end, she analyses the ways in which the structural adjustment programme advocated by the government of Peru (at the behest of the international financial community) affects the economic, social and cultural rights of women in that country. She compares the differential experiences of economic restructuring on women and men working in the informal sector. Hays-Mitchell first examines the human rights dimension of economic restructuring by exploring the relationships between structural adjustment and the feminisation of poverty, considering especially the influence of cultural and social norms in determining women's access to productive resources and, hence, gainful livelihoods. Then she examines gender-specific responses to the disenfranchising conditions of restructuring by focusing on the grassroots initiatives of women informal workers to overcome their exclusion from conventional sources of finance capital and technical training. Her analysis and discussion is based on fieldwork conducted in 1993–4 in Lima, Peru. The case study illustrates how the exclusion of women from programmes for micro-enterprise development in Peru limits their ability to access productive resources and thus to participate equally in the process of development. This exclusion constitutes a violation of the basic human rights of women to development as set forth in the 1986 United Nations Declaration of the Right to Development. Moreover, it uncovers the structural discrimination on which economic restructuring in Peru is premissed.

In Chapter 8, Eleonore Kofman discusses the rights of family reunification and formation during migration and the implication for some of the crucial policy areas of employment and housing in the European Union. These policies have proved to be major obstacles for women migrants even more than for men. Their implications are particularly significant, since family reunification and formation currently account for the largest flow of legal migrants into Europe and their applications for migrant status are subject to stringent controls. For women, family-related migration has been by far the dominant mode of entry, whether they enter as wives and fiancées, or themselves apply to bring in family members. In each case, they encounter particular problems in meeting the regulations prescribed by states. Kofman first outlines the changing nature and context of immigration in the 1980s and 1990s in Europe with particular emphasis on women. She then examines the coverage of international conventions in relation to migrant rights, the experiences of women in family reunion and the ways in which changing regulations are likely to effect their ability both to enter as family migrants and to bring in family members.

In Chapter 9, Deborah Rose examines the evidence with respect to the human rights of Indigenous people in Australia ; she describes an Indigenous system of gendered power and knowledge as a basis for analysing cultural survival. In addition, she analyses some of the contexts relating to land rights in which Indigenous women have been marginalised. She reflects upon some questions of the relationship of land rights to the promotion of human rights via issues of cultural survival, gender equity, and social equity. Her basic premiss is that human rights is about empowerment; she argues that in the context of land rights, central to Indigenous people's culture, the colonising society imposes a set of gender constructs that marginalises and disempowers Indigenous women. She analyses how lack of consideration of cultural survival affected Indigenous women in two planning projects: the Alice Spring Dam project and a bridge-building project in South Australia.

The final contribution, by Jiřina Šiklová, explores the ways in which 'human rights' were conceptualised under the socialist regimes in Eastern Europe and comments on reasons why women have not taken up the call to associate 'women's rights' and 'human rights' in the post-socialist Czech Republic. She explores two areas in more depth, women's rights in employment and women's rights in relation to their bodies, identifying issues that activists and planners need to address if women's human rights are to be better realised within the context of a post-Communist state.

With Jiřina Šiklová's chapter on women's human rights and their bodies, a wide variety of human rights analysis in planning in different geographies and spatialities comes to its end. It is hoped that in offering these diverse contributions the book will form the foundation of a new agenda for planning which will be more sensitive to gendered human rights and in the process, recognise the diversity of women.

NOTES

1 Rational comprehensive methodology of planning consists of several logical levels. These start with problem definition and develop through data collection and processing, the formation of goals and objectives and the design of alternative plans. Finally, there are the processes of decision-making, implementation, monitoring and feedback (Moser 1993).
2 The 'blueprint' approach includes surveying, analysis and planning.

REFERENCES

Alterman, R. (1994) 'The theoretical bases for planning and its implications in defining plan's objectives and goals', in *Master Plan for Israel – 2020*, Tel Aviv (in Hebrew).
Amnesty International (1995) *Human Rights are Women's Rights*, Amnesty International.
An-Na'Im, A. (1995) 'State responsibility under international human rights laws to change religious and customary laws', in R. Cook (ed.) *Human Rights of Women*, Philadelphia, PA.: University of Pennsylvania Press.
Binion, G. (1995) 'Human rights: a feminist perspective', *Human Rights Quarterly* 17(3): 509–26.
Bunch, C. (1995) 'Transforming human rights from a feminist perspective', in J. Peters and A. Wolper (eds) *Women's Rights, Human Rights*, New York: Routledge.
Charlesworth, H. (1995) 'What are "Women's International Human Rights"?', in R. Cook (ed.) *Human Rights of Women*, Philadelphia, PA.: University of Pennsylvania Press.
Cook, R. (1995) 'Women's international human rights law: the way forward', in R. Cook (ed.) *Human Rights of Women*, Philadelphia, PA.: University of Pennsylvania Press.
Coomaraswamy, R. (1993) 'The principle of universality and cultural diversity', in M. A. Schuler (ed.) *Claiming Our Place: Working the Human Rights System to Women's Advantage*, Washington, DC: Institute for Women, Law and Development.
Coomaraswamy, R. (1995) 'To bellow like a cow: women, ethnicity, and the discourse of rights', in R. Cook (ed.) *Human Rights of Women*, Philadelphia, PA.: University of Pennsylvania Press.
Eisenstein, Z. (1996) 'Women's publics and the search for new democracies', paper presented at the conference: Women, Citizenship and Difference, in London, 16–18 July, 1996.
Fainstein, S. S. (1996) 'Planning in a different voice', in S. Campbell and S. Fainstein (eds) *Readings in Planning Theory*, Oxford: Blackwell Publishers.
Foucault, M. (1977) *Power/Knowledge*, New York: Harvester Wheatsheaf.

Friedman, J. (1996) 'Feminist and planning theories: the epistemological connection', in S. Campbell and S. Fainstein (eds) *Readings in Planning Theory*, Oxford: Blackwell Publishers.

Gasper, D. (1996) 'Culture and development ethics: needs, women's rights, and western theories', *Development and Change* 27(4): 627–61.

Gomez, M. (1995) 'Social economic rights and human rights commissions', *Human Rights Quarterly* 17(1): 155–69.

Howard, R. (1995) 'Women's rights and the right to development', in R. Cook (ed.) *Human Rights of Women*, Philadelphia, PA.: University of Pennsylvania Press.

Huxley, M. (1994) 'Planning as a framework of power: utilitarian reform, enlightenment logic and the control of urban space', in S. Feber, C. Healy and C. McAuliffe (eds) *Beasts of Suburbia*, Melbourne: Melbourne University Press.

Ley, D. (1989) 'Fragmentation, coherence and limits to theory in human geography', in A. Kobayashi and S. Mackenzie (eds) *Remaking Human Geography*, London: Unwin Hyman.

Lister, R. (1995) 'Dilemmas in engendering citizenship', *Economy and Society*, 24(1): 2–40.

Marsall, T. H. (1950) *Citizenship and Social Class*, Cambridge: Cambridge University Press.

Massey, D. (1994) *Space, Place and Gender*, Cambridge: Polity Press.

Moore, H. (1996) *Space, Text, and Gender*, New York: Guildford Press.

Moser, C. O. N. (ed.) (1993) *Gender Planning and Development: Theory, Practice and Training*, London and New York: Routledge.

Nussbaum, M. (1995) 'Human capabilities, female human beings', in M. Nussbaum and J. Glover (eds) *Women, Culture and Development*, Oxford: Clarendon Press.

Paris, C. (1982) 'Introduction', in C. Paris (ed.) *Critical Readings in Planning Theory*, London: Pergamon Press.

Pateman, C. (1988) *The Sexual Control*, Cambridge: Polity.

Pateman, C. (1989) *The Disorder of Women*, Cambridge: Polity.

Rao, A. (1995) 'The politics of gender and culture in international human rights discourse', in J. Peters and A. Wolper (eds) *Women's Rights Human Rights*, New York: Routledge.

Ritzdorf, M. (1996) 'Feminist thoughts on the theory and practice of planning', in S. Campbell and S. Fainstein (eds) *Readings in Planning Theory*, Oxford: Blackwell Publishers.

Safier, M. (1990) 'Making plans and making cities: creating, recognizing and sustaining a room for manouver in planning practice', mimeo.

Schuler, M. A. (ed.) (1993) *Claiming Our Place: Working the Human Rights System to Women's Advantage*, Washington, DC: Institute for Women, Law and Development.

Sullivan, D. (1995) 'The public/private distinction in international human rights law', in J. Peters and A. Wolper (eds) *Women's Rights Human Rights*, New York: Routledge.

Valentine, G. (1989) 'The geography of women's fear', *Area* 21(4): 385–90.

Wetzel, J. W. (1993) *The World of Women in Pursuit of Human Rights*, Basingstoke and London: Macmillan.

Young, K. (1993) *Planning Development with Women*, London: Macmillan.

Yuval-Davis, N. (1997) *Gender and Nation*, London: Sage.

Part II

GENDER, PLANNING AND HUMAN RIGHTS

2

WOMEN, PLANNING AND LOCAL CENTRAL RELATIONS IN THE UK

Jo Little

INTRODUCTION

In this chapter broad questions of women's rights are translated to the local level in the analysis of the town planning process in the UK. The chapter reviews the progress that has been made in confronting gender inequality in the built environment through town planning legislation and practice. While much existing work in this area (and indeed, the focus of this chapter) is on the detail of town planning policy and its implementation, it is important to recognise at the outset the wider connections to women's rights and the relevance of the more global perspectives adopted by many of the other chapters in this book. Town planning in developed economies clearly has the ability to influence key aspects of everyday life in which women's human rights are frequently abused – aspects such as accessibility to resources, employment and housing which underlie the essential opportunities and life chances experienced by women.

Central to understanding the relationship between women's rights and town planning is the notion of citizenship and the belief that aspects of planning legislation as well as the overall direction of planning policy serve to deny the capacity of women to influence their own lives and to play a role in decisions about resource allocation. As Walby (1997) argues 'citizenship', is imbued with gender-specific assumptions which relate to the organisation and use of space and resources and which restrict women's ability to participate fully in the public realm. Women's lack of power over the planning system is, as shown below, critical not only to their position within the local state but also to their broader access to citizenship. Powerlessness underlies the abuse of women's human rights in all spheres – powerlessness in the arena of local and national planning represents an attack on women's basic rights and on their access to citizenship.

Within this wider concern for citizenship and women's rights the chapter begins by addressing strategic questions concerning the status of and commitment to the use of women's initiatives in planning as a means of reducing gender inequality. Drawing on local examples, the chapter then goes on to place planning's lack of success in responding to women's inequality in the context of the day-to-day organisation and operation of town planning departments and to the dominance of male control of the decision-making system. Finally links are made between the current position of women's issues in planning and the operation of feminist politics in the local state. This chapter shares with the other chapters a desire to provide an explicitly feminist analysis of the operation and outcome of the planning system. Striving to apply a feminist perspective also unites the detailed local level analyses with the more global picture and again helps to make sense of the specific material included here in the context of a broader human rights perspective. It is possible, for example, to see how local patterns in gender inequality and women's subordination have their origins partly in wider (and sometimes global) systems of power and resource allocation.

The second half of the 1980s saw the publication of a number of articles and reports on the subject of 'women and planning' in the UK. The majority of such publications remarked on the absence of town planning initiatives aimed at so-called 'women's needs'. This they linked to a lack of research into the particular needs of women within the built environment and the degree to which women are disadvantaged within the conventional town planning process. While town planning has certainly been slow to respond to the growing awareness of women's needs and gender difference in the built environment it would be wrong to assume that there has been no change from this earlier position. Today there is a greater and more broadly based awareness of the differing needs of men and women within the planned environment and a more widespread acceptance (both within the academic but also the practising planning community) of the legitimacy of identifying and responding to the particular needs of women.

There now exists a number of academic reviews of the direction and contribution of policies or initiatives aimed at women's needs (see, for example, Foulsham 1990; Greed 1994; Little 1994a). In addition, a growing number of local authorities have published documents outlining the responses they have made to the problems seen to be faced by women within the built environment (see, for example, Islington Women's Equality Unit 1991; Leicester City Council 1991; Southampton City Council 1991). The work that has been done has gone some way towards establishing the nature of women's disadvantage within the built environment and the potential of planning to respond to issues of gender inequality.

Despite this growing awareness amongst academic and practising planners in the UK, there still remains a major gap. This, I would argue, is because our understanding of the experiences of women in the built environment and of the failure of planning to respond to women's needs exists in a vacuum. To this end it continues to be largely untheorised and certainly outside any broader consideration of male power either *within* the planning process or as part of the organisation and operation of the local state. It also, just as critically, lacks an appreciation of the differences between women in terms not only of the content of individual initiatives but women's relationship with both the national and local power structures. Both weaknesses hinder attempts to see women's powerlessness in planning as a more basic attack on their status and political rights in society. This chapter, therefore, focuses on the neglected issue of gender and power in planning and in so doing adopts, as noted above, an explicitly feminist analysis. Questions of difference *between* women are relevant throughout the discussion. They do not form a central line of analysis here, however, owing to the constraints of time and to the need to focus on the central issue of power relations between men and women in relation to individual rights.

WOMEN'S INITIATIVES IN TOWN PLANNING

As already indicated, a number of planning authorities have begun to develop initiatives aimed at addressing women's needs within the planned environment. In general, these have mainly conformed to a set pattern. They are largely concentrated within particular topic areas – employment, transport, childcare, service provision and safety – and reflect a standard interpretation of both women's needs and the sorts of practical planning responses that can be made. Policies include, for example, a requirement that new retail development over a particular size should incorporate childcare facilities or that street design be sensitive to women's safety in terms of lighting, vegetation and car parking etc. While I do not wish to undermine or dismiss such initiatives, what is clear is that their introduction has reflected a very conventional response. In addition, those initiatives that have been attempted have been mainly isolated and individual reactions to particular perceived problems rather than part of a broader based strategy to increase women's rights.

The extent and nature of planning initiatives as introduced by local authorities in England has been examined in some detail elsewhere (see Little 1994a, 1994b). A review of all local authority planning departments in the early 1990s[1] revealed that 23 per cent had introduced some form of women's initiative through formal planning policy. This review allowed the development of a simple typology for categorising the departments' responses to women's needs and the possible introduction of women's initiatives, as listed below.

1 An awareness of women's needs appears to have penetrated into many or all policy areas. Individual initiatives are part of a wider commitment to promotion of equality within the built environment.
2 Response to problems faced by women is on an *ad hoc* basis. Specific initiatives are 'added in' to existing policies individually. No broader strategy for tackling gender inequalities exists within the department and there is no obvious shift in broader priorities to reflect women's needs.
3 There is no awareness of the nature of women's needs in the built environment or there is a denial that planning problems and solutions are gendered. No recognisable attempts are made to introduce 'women's initiatives'.

Despite the generally higher profile of 'women's initiatives' in planning described earlier, the majority of authorities reviewed fell into the third of these categories, remaining unconvinced by either the need for or the value of women's initiatives as part of the planning process. Where authorities had introduced initiatives, the majority, thirty-three, had done so in just one policy area (and hence accorded with category 2 of the typology). Only seven authorities had introduced initiatives in three or more policy areas and could claim to demonstrate a widespread commitment to addressing women's inequality through policy initiatives.

It is not the purpose here to present a detailed analysis of the conclusions of the earlier research; other general points such as the concentration of women's initiatives in urban authorities and in authorities controlled by councillors belonging to the Labour Party, reinforced conclusions drawn by other studies (see Brownill and Halford 1990; Halford 1989). The intention rather is to consider some of the reasons for the continued reluctance of planning departments to develop women's initiatives and to do so with particular reference to power relations within the planning process. First, however, it is useful to place the above discussion of local level planning policy in the context of the broader status of women's initiatives in planning nationally.

In 1995 the Royal Town Planning Institute (RTPI) published a Practice Advice Note (PAN) setting out what is considered 'good practice' in relation to (in this case) women and planning. This PAN has no legislative status but is seen as providing guidelines to be used by planners in the course of both day-to-day development control and in more strategic plan-making activities. The PAN entitled 'Planning for Women' aims to

> clarify how policies can *unwittingly discriminate* against groups such as women and how members of the Institute (the RTPI) can use their influence in the public, private and voluntary sectors to promote good planning practice sensitive to the needs of women.
>
> (RTPI 1995: 1, my emphasis)

The PAN outlines aspects of legislation and government guidance in relation to sexual discrimination. It goes on to quote the Department of the Environment good practice guide for development plans which concluded that

> women should be *added in* to the list of particular groups whose needs should be taken into account in development plans.
>
> (RTPI 1995: 12)

This PAN provides, it may be argued, a good illustration of the general direction of 'planning initiatives for women' in all but a very small minority of authorities. It is characterised by a few key directions:

- it covers only very conventional and 'accepted' areas of planning for women;
- it is highly design centred;
- it is at pains to point out that planning 'for' women benefits the wider community in that it emphasises that 'shaping towns and villages that meet

28

women's needs can help to create an environment that works better for *everyone* in society' (RTPI 1995:1);

- the tone is at times apologetic with the emphasis on the addition to rather than alteration of existing policy.

The PAN, then, offers nothing new in terms of the accepted views of women's inequality within the built environment. More importantly, perhaps, it contains no real challenge to the existing priorities of planning policy and the planning process. There is no recognition of the way in which the built environment reflects the unequal distribution of power between women and men or that the existing form of the built environment and direction of planning serves to further certain (usually male-dominated) interests above others. Clearly the direction and tone of the document is influenced by its purpose (as an advice note) but the total avoidance of discussions of gender relations and inequality may be seen as indicative of the document's limited scope.

The most progressive point made by the PAN is that there is a need to encourage women within the decision-making process. It does not go as far as saying that this process needs to be revised so as to facilitate this involvement, however. It also makes the point that the needs of all women will not be identical. Again this is not followed by any advice on how the differing needs of women may be recognised and addressed through policy and practice.

The PAN, together with the discussion of research at the local level as presented earlier, provides a useful indication of the limits to the recognition and amelioration of inequality based on gender within the UK planning system and its implementation. Obviously it is not comprehensive but it does serve to demonstrate the lack of a clear feminist agenda in contemporary planning. Disappointingly the academic discipline of planning, while giving attention to the need for greater emphasis to be placed on 'women's needs', has not pushed such an agenda either through its commentary on existing planning directions or in its vision for the future.

As the first stage in the development of a feminist agenda for planning it is essential that we start to look critically at the reasons for planning's lack of commitment to addressing gender inequality. To do this we need to consider the internal power relations within the planning process, the patriarchal attitudes of those in positions of power as well as the actual systems and processes that exist for handling gender issues. It is to these issues that the chapter will now turn.

WOMEN AND DECISION-MAKING WITHIN THE LOCAL AUTHORITY

This section will look first at the formal structures which exist in local authorities in the UK for addressing women's rights and issues of gender inequality. The aim here is to consider the relationship between the status of and commitment to women's initiatives in planning and the more general authority-wide mechanisms for confronting gender inequality. The section examines, in particular, the extent to which women's committees within local authorities can stimulate more positive approaches to women's needs in the context of planning policies. Again it stresses the link between meeting the day-to-day needs of women and increasing their rights, showing how access to power and to broader notions of representation is central to the detail of women's daily lives.

Local authority women's committees were introduced in the early 1980s specifically to address the perceived needs of women at the local level. They were generally authority-wide organisations with a remit to advise other departments on relevant issues and to work with them to identify problems and solutions. Some had substantial budgets when they were established (Bristol City Council Women's Committee, for example, was allocated a budget of £300,000 in its first year of existence). What is most relevant here is not the actual achievements *per se* of those women's committees that were set up but more the pattern of their introduction, growth and decline as a part of local government and how this relates to both the internal and external politics of the local authority, and particularly to the state of the New Urban Left in the 1980s.

Perhaps the most significant feature of the history of women's committees in local government is their relatively short lifespan. Table 1 shows this decline over a

Table 1 Women's committees in local authorities

Type of organisation	1986	1991	1995
Full committee	14	12	6
Sub-committee	8	5	5
'Other' initiative	9	18	n.d.

n.d., Data not available
Sources: Halford 1989; Questionnaire survey 1991, as referred to in note 1; Municipal Year Book 1995, HMSO.

period of nine years. What is also noticeable from this table is the increasing importance of 'other initiatives' for addressing women's needs. These other initiatives are generally less formal, often with no financial or policy-making status in the authority. They range from semi-formal groups set up to advise on or respond to particular issues in relation to women's needs locally, to much more *ad hoc* groups often providing a support function for women working within the authority rather than as a wider input into policy-making.

Research undertaken in 1991 indicated a clear link between the existence of a women's committee within a local authority and the likelihood of that authority's planning department introducing women's initiatives; 80 per cent of authorities with a women's committee (or sub-committee) had introduced planning initiatives aimed at women's needs (the relationship between planning and the 'other' initiatives was less clear). Consequently the decline in the number of local authority

30

women's committees over the nine years indicated will potentially have serious ramifications for planning's treatment of gender inequality.

To explore, in more depth, this relationship between the existence of women's committees and planning initiatives for women, I interviewed a number of planners working in Bristol City Council (a large urban authority in the south-west of England), most significantly the Head of Local Plans team. The conclusions from this study may not be representative of all authorities but they do help to shed some light on the direction of the relationship between planning and women's committees within an authority with a generally positive record and attitude towards addressing inequality.

Bristol City Council and the changing relationship between planning and the Women's Committee

Bristol City Council set up a Women's Committee in 1987. This committee was supported by a full-time Women's Unit with a paid staff of three. As indicated above, the committee was established with a substantial budget. In 1990, after just three years of operation, a review of the whole authority by consultants recommended the abolition of the Women's Committee as a full committee and its merger with the Race and Disabilities Committees to form an 'Equalities Committee'. This reconstitution was followed by the disbanding of the Women's Unit, and with it the removal of an important resource for local women. The politics behind these moves is discussed in the final section of the chapter; here we are concerned with the direct and identifiable implications for planning.

Overall, the view of planning officers at Bristol City Council was that the abolition of the Women's Committee as such *had* led to a clear downgrading of women's issues within planning. Most noticeably, according to the Head of Local Plans, less pressure was now placed on planning by the new Equalities Committee to ensure that women's interests had been considered in the formulation and implementation of policy. The past chair of the Women's Committee had been 'very challenging' and kept officers 'on their toes'. Now, with less direct pressure, the Head of Local Plans admitted that she 'listened to only about 25 per cent' of what was said in relation to equality issues. Pressure on planning officers to respond to the Equalities Committee has also eased by the removal of formal links (e.g. a designated officer and regular meetings such as existed in the past between planning and the Women's Committee). Papers were 'shunted through' to planning on a more *ad hoc* and informal basis.

In terms of the reality of planning decisions and the evolution of policy, it is possible, then, to identify a 'watering down of initiatives aimed at meeting women's needs'. The Local Plan produced by Bristol City Council in 1993, for example, made little reference to the specific needs of women, pursuing instead, so the Head of Local Plans claims, an 'equalities thread'. Her argument was that the problems experienced by women did not go unrecognised but that these problems were best addressed through the promotion of 'good planning'. As in

31

the previously discussed Planning Practice Advice Note, planning for women was seen in Bristol to be a case of 'planning for all' and not 'discriminating' against or in favour of *any* particular group. The retreat of planning from a direct (and potentially contentious) concern with women's inequality and its reluctance to challenge the status quo (and thus challenge the view that women's needs will automatically be met through the promotion of fairness) is something which is seen to be linked, in part at least, to a reduction in the internal pressure exerted by the presence of the Women's Committee.

The downgrading of the Women's Committee in Bristol can be interpreted, in isolation, simply as an attempt by the public sector to re-assess equality issues in general. Alternatively it can be interpreted as a very real attack on the ability of women in the city to influence or control decisions made over public resources. Seen in this broader context it is clearly of direct relevance to the position of women in society and to the gender division of power as it relates to basic human rights.

Any analysis of the commitment of planning to reducing women's inequality in the built environment must also look beyond the more formal structures of power to the broader cultures of decision-making within authorities and departments. Clearly in cases where the formal mechanisms for addressing inequality are absent, weak or changing and in situations where little exists by way of written guidelines, the planning decisions that are made over issues relating to women's needs may be very significantly influenced by the particular commitment of individuals and the general attitudes and priorities prevailing within departments and authorities. The next part of this section will consider some of these less formal aspects of the policy environment as they relate to the treatment of women's inequality by planning departments.

Planning for women and cultures of decision-making

A major influence on the willingness of planning to respond to women's inequality in the built environment is, so researchers have argued (see Davies 1993; Greed 1992), the position of women within the planning profession. Work such as that by Greed has pointed to the absence of women as qualified planners, especially at the more senior levels, as a key factor in planning's failure adequately to address gender inequality in the built environment. While it may not be possible to draw a direct correlation between women's involvement in planning as professionals and the introduction of women's planning initiatives, it can be argued that women's presence as planners is important in raising awareness of women's experiences within the built environment generally and of appropriate responses to their needs. Conversely, the absence of women from the planning profession can result in a lack of sympathy towards women's needs and is more likely to encourage a retreat to the status quo.

In the course of research referred to earlier (see Little 1994a), I collected comments from planning officers on their attitudes to planning's role in addressing gender inequality and women's needs. Within the general response to my questions about the importance of 'women's needs' within the authorities policies, the following comments were not untypical:

Land use planning policies are formed in relation to development pressures and proposals, in order to control/guide physical development. In 99.9 cases out of 100 the needs of a particular sex are not a relevant factor in consideration of proposals.

(Town Council Planning Officer)

I don't regard 'women's needs' to be any different to men's.

(District Council Planning Officer)

Both men and women are regarded as people, just as all races are. We don't discriminate positively or negatively.

(District Council Planning Officer)

Other comments related to the links between an authority's record on adopting women's initiatives and the involvement and status of women in the planning profession.

(There is) very little appreciation of how planning, especially development control (DC), can influence life for women in the community. There are only two women DC officers and little encouragement to undertake initiatives to aid women. Major developments where crèche or pram parks etc. could be incorporated tend to be dealt with by senior male officers.

(Borough Council Planning Officer)

Unfortunately anything that comes in about it (women and planning) is handed to a woman to sort out. There are no women on the council's management team and women's issues are not addressed, managed or anything else. Perhaps filed.

(District Council Planning Officer)

I am sure there are too few women councillors – reflecting the bias throughout the whole of the planning system. If there's no support from politicians or from senior management it is difficult to implement new policies aimed at women.

(County Council Planning Officer)

Discussion with planning officers at Bristol City Council provided more detail on these general comments on both the links between the appointment of women to senior positions within an authority and the priority given to women's needs and to the persistence of some very stereotypical attitudes towards women's initiatives within the broader culture of planning offices.

Bristol City is regarded as having a good record on appointing women staff and it was felt by those working in the department that there was a general sympathy amongst staff towards women's needs. The point was made, however, that it is *still*

33

the case that the majority of senior staff (in planning and, crucially, in other departments) are male and that traditional attitudes towards women in the work-place continue to exert themselves. As one senior woman officer argued, although the CPO (Council Planning Officer) for the council is herself a woman (and one with a reputation for supporting women's interests), she has 'a top job in a man's world'. In this world she is judged by her peers and has only succeeded in getting where she is by adopting their approach.

While Bristol City may be seen as a left of centre authority that is sympathetic to the needs of 'minority groups', in planning terms the 'kudos', according to the Head of Local Plans, still lies in environmental and economic interests. There is a recognition by the CPO and her staff that reputations won't be made or broken on 'women's issues' but on the mainstream environmental and economic issues. Two recent examples were given where planners were not prepared to 'stick their necks out' for women's interests over the more traditional direction of policy. The first concerned a major new shopping development in the middle of the city and the developer's reluctance to include a shoppers' crèche and to fund toilets with baby changing facilities. The second was the allocation of new 'docklands' hous-ing not for low cost 'social' housing but for conventional 'yuppie' development. In both cases the risk of losing the proposed development was seen as too important to allow the more 'peripheral' interests to be pursued.

These examples from Bristol demonstrate the fact that even where individual women have gained senior positions within the decision-making hierarchy, it is often impossible for them to fight against existing trends. As the Head of Local Plans at Bristol stressed, a lack of support at the top makes it impossible for women to 'put their heads on the block and stand up for a greater con-sideration of women's needs'. The continued existence of traditional attitudes and priorities is reinforced, moreover, by a set of working practices which, again, are generally too entrenched to be challenged by individuals. At Bristol, a declared concern with women's interests and equal opportunities in the workplace are somewhat overshadowed by an absence of crèche facilities in the council, the continuation of a practice of evening meetings (especially common for senior staff), poor communication and sexist attitudes towards the reliability of working mothers.

There is insufficient space here to consider these issues in more detail. There are many aspects of workplace culture which could be examined in relation to women's issues; the way in which meetings are held, for example, or the provision and management of training for council employees. The point has been made, however, that both the position of women within the planning profession and the working environment of planning officers are highly influential on the adoption of planning initiatives aimed at addressing women's needs and on the wider sympathy towards the importance of women's issues within planning. This rela-tionship is not straightforward and is often difficult to identify. It is essential, however, that despite this difficulty it is recognised as relevant to questions of gender inequality in the built environment and to any discussion of ways of improving women's experience through the planning process. This chapter now turns finally to set the review of planning and women's initiatives within the context of local politics and the operation of the local state. In so doing it will return to questions of power in a wider sense as they relate to gender inequality and the feminist political agenda.

LOCAL POLITICS AND PLANNING INITIATIVES FOR WOMEN

Recent work in Britain on the changing face of local politics has drawn particular attention to the attack on the local state by successive Conservative governments and the corresponding decline of the New Urban Left, as it became known (see, for example, Cloke 1992; Duncan and Goodwin 1988; Hambleton and Thomas 1995). The effects of this decline from the mid-1980s have been detailed as they have been experienced by different groups in the urban environment, with particular emphasis being placed on the impact on the politics of consumption. While the early 1980s witnessed a considerable buoyancy in the profile of and involvement in campaigns about consumption issues, the late 1980s saw the increasing fragmentation of local groups and the breaking down of the sense of common purpose that characterised earlier local community action. Intense local activity was replaced by a small number of national campaigns, which have been constituted outside the domain of local politics and, at the same time, successfully portrayed as marginal or 'other' by the Conservative government. Examples include campaigns on the introduction of the poll tax, the rights of new age travellers and protests about animal rights.

As has been noted elsewhere (see, for example, Chouinard 1996; Coote and Pattullo 1990) campaigns on consumption issues have traditionally involved women disproportionately, providing the route by which many have entered local and even national politics. Consequently, the downgrading of such campaigns and of the role of the local state generally was particularly damaging to women's role in political activity and to local feminist action. The feminist movement was hit hard by the fragmentation of political activity and by divisions which formed along the lines of class and race. In keeping with other groups, the cohesion of local women's groups was reduced, with an inevitable effect on their power and influence. Conservatives at both the local and national level sought further to fragment local action in the portrayal of left-wing groups (especially feminist groups) as 'loony' and somehow outside 'reasonable' political activity.

Brownill and Halford (1990) have linked the establishment of women's initiatives in local government (both formal structures such as women's committees and individual policy initiatives) to the particular influence of local feminist activity.

> Evidence suggests that the reason why only certain local authorities chose to set up *women's* initiatives arises from variations in local gender relations,

which in turn are connected to variations in feminist political activity and ultimately in distinctive local policy outcomes.

(Brownill and Halford 1990: 411, original emphasis)

In this way it is also possible to relate the *decline* in women's initiatives to local gender relations and to a decline in feminist activity at the local scale. Similarly, the reluctance of planning departments, as a function of local state activity, to address the needs of women in their decision-making can be seen, in part at least, to be a result of the changing power and influence of women in local formal and informal politics. It also relates to the broader position of left-wing politics locally and to feminist activity within the wider context of local socialism.

However, identifying an immediate or causal relationship between women's initiatives and feminist or left wing politics is essentially problematic. The conclusions of authors such as Brownill and Halford are based largely on the direction of policy and of political activity rather than the specifics of any particular initiative or change. As with earlier arguments concerning the relationship between formal structures of decision-making and planning initiatives, a more detailed local level analysis can help to establish more clearly the basis of the tendencies and trends that exist. Hence the final part of this section will return to the case of Bristol in order to take up some of these issues.

A full understanding of the relationship between women's initiatives and local politics in Bristol City would require a much more comprehensive analysis of the history and structure of political decision-making than can be provided here. As it is there are some easily identifiable changes that can be seen as having a very clear impact on that relationship and indicative of the overall direction of local state activity.

Changes in the local Bristol economy during the late 1970s and 1980s manifested themselves in the social and political structure of the city such that Bristol increasingly became a middle-class professional city embracing the so-called 'new municipal socialism' of the GLC and other new left councils (Bassett 1996). The traditional base of Labour Party support was replaced by a younger more middle-class group, but although many individuals from this group gained positions of authority within the city council, they failed to gain overall control of the Labour group within the council. The result was that infighting was common over some of the key areas of radical concern in the 1980s (notably feminist and racial issues). As Brownill and Halford (1990) observed more generally, this infighting was particularly critical over the co-option of members of the Women's Committee (the working-class and lesbian members specifically) and reflected deep divisions between factions of the feminist movement in the city as a whole.

The Women's Committee became an important site for conflict between the opposing factions of the Labour Party in the city council. To some extent this mirrored wider divisions in local political activity but was also responsible, itself, for perpetuating those divisions. A lengthy campaign to discredit the Women's Committee (with accusations of overspending, unfairness and the pursuit of marginal interests) was entered into by the old style Labour members and Conservative councillors alike and led eventually to internal divisions within the committee. This was instrumental in the decision to downgrade the Women's Committee (and abolish the Women's Unit) and to introduce an Equalities Committee, as discussed earlier.

During the late 1980s, feminist activity within the city was declining. Divisions emerged within established women's groups – often along the lines of class – reflecting the tensions between 'old style' working-class and the new middle-class women. Particularly relevant in this context is that no comments were received from women's groups on the most recent version of the city's Local Plan. Initial concern from women's groups over the disappearance of the Women's Committee was not built upon, and many women accepted the replacement of the Women's Committee with the Equalities Committee as the only solution to class divisions and the factional infighting that was characterising both the city council and local feminist activity. Separating the formal and the informal to suggest the 'cause' of the decline in feminist politics and the reduction in the emphasis on women's interests within the built environment is impossible, and changes in each sphere can be seen as acting in a mutually reinforcing way.

CONCLUSION

This chapter has provided an overview of the current state of women's initiatives in planning in the UK. The aim has been to depart from the traditional (and very restricted) focus of accounts of 'women and planning' to provide a more critical analysis of the failure of planning departments to promote the cause of women's rights. This analysis has attempted to explain the failure by reference to the operation of power within the decision-making process and to the wider relationship between planning, the local state and local political activity. The importance has been recognised of locating the study of women and the planning system within an understanding of the links between the planning process and human rights. By focusing on the underlying power relations within the planning system it has shown how a failure to meet women's needs in the built environment relates to gendered assumptions about citizenship and the rights of individuals in contemporary society.

To return to the central focus of women's rights: a feminist analysis shows that the contemporary planned environment not only reflects gender inequality but also serves to reinforce (and perhaps even to produce) that inequality by making it difficult for women to participate equally in all spheres of life (MacGregor 1995). A feminist critique of planning seeks to explain the current (and enduring) form of the built environment by reference to the forces that shape decision-making. It is only as we come to acknowledge the patriarchal nature of policy-making, including the culture of the decision-making process, and its roots in the wider operation of gender relations in the local state that we can fully appreciate the basis of women's inequality and work towards a more equitable form of the built environment.

NOTES

1 In 1990 and 1991 I contacted all local authority planning departments in England. A questionnaire was sent to each authority in which basic questions about the number and content of planning initiatives for women and broader questions about gender inequality and training were asked. A response rate of over 65 per cent was achieved on the questionnaire. In a few cases this was followed up by longer one-to-one interviews with planners.

REFERENCES

Bassett, K. (1996) 'Partnerships, business elites and urban politics: new forms of government in an English City?', *Urban Studies* 33(3): 539–55.
Bristol City Council (1993) *Local Plan*, Bristol: Bristol City Council.
Brownill, S. and Halford, S. (1990) 'Understanding women's involvement in local politics: how useful is a formal/informal dichotomy?', *Political Geography Quarterly* 9(4): 396–414.
Chouinard, V. (1996) 'Gender and class identities in process and in place', *Environment and Planning A* 28(8): 1485–506.
Cloke, P. (ed.) (1992) *Policy and Change in Thatcher's Britain*, Oxford: Pergamon Press.
Coote, A. and Pattullo, P. (1990) *Power and Prejudice: Women and Politics*, London: Weidenfeld and Nicolson.
Davies, L. (1993) 'Aspects of equality', *The Planner* 79(3): 14–16.
Duncan, S. and Goodwin, M. (1988) *The Local State and Uneven Development*, London: Polity Press.
Foulsham, J. (1990) 'Women's needs and planning: a critical evaluation of local authority practice', in J. Montgomery and A. Thornley (eds) *Radical Planning Initiatives: New Directions for Urban Planning in the 1990s*, Aldershot: Gower.
Greed, C. (1992) 'Women in planning', *The Planner* 78(13): 11–13.
Greed, C. (1994) *Women and Planning: Creating Gendered Realities*, London: Routledge.
Halford, S. (1989) 'Spatial divisions and women's initiatives in British local government', *Geoforum* 20(2): 161–74.
Hambleton, R. and Thomas, H. (eds) (1995) *Urban Policy Evaluation*, London: Paul Chapman.
Islington Women's Equality Unit (1991) *Islington Women's News*, London: London Borough of Islington.
Leicester City Council (1991) *Local Plan*, Leicester: Leicester City Council.
Little, J. (1994a) *Gender, Planning and the Policy Process*, Oxford: Pergamon Press.
Little, J. (1994b) 'Women's initiatives in town planning in England: a critical review', *Town Planning Review* 65(3): 261–78.
MacGregor, S. (1995) 'Deconstructing the man-made city: feminist critiques of thought and action', in M. Eichler (ed.) *Change of Plans: Towards a Non-Sexist Sustainable City*, Toronto: Garamond Press.
Royal Town Planning Institute (RTPI) (1995) *Planning For Women*, Practice Advice Note 12, London: RTPI.
Southampton City Council (1991) *Women and the Planned Environment: Design Guidelines*, Southampton: Southampton City Council.
Walby, S. (1997) *Gender Transformations*, London: Routledge.

3

CULTURE, HUMAN RIGHTS AND PLANNING (AS CONTROL) FOR MINORITY WOMEN IN ISRAEL

Tovi Fenster

INTRODUCTION

This chapter explores the dilemma of whether to take cultural values into consideration in spatial planning, even if these values abuse women's human rights. It focuses on the planning processes undertaken for two minority communities in Israel, the Muslim Bedouin and the Ethiopian Jews, both undergoing cultural and social change.

A point of clarification is necessary at the outset. The debate raised in this chapter concerns 'top-down' planning, that is, planning by professionals, usually from different cultures from the people for whom they are planning. Obviously, 'bottom-up' or self-made planning by the people themselves is desirable, but bearing in mind the reality in the world in which probably most planning schemes are made by 'outsiders', it seems relevant and constructive to consider a 'top-down' dilemma regarding planning.

In light of the above, the main aim of the chapter is to highlight the problematic situation of 'top-down' planning for ethnic minorities. Should cultural norms perpetuating women's subordination be regarded as serving a 'human need' that planning schemes have to take into consideration; if not, then to what extent should planning schemes play a role in changing these norms? In other words, should planners take a 'western universalist' or 'local relativist' view on cultures in the planning process? The former means that plans must be made according to the principles of one generic western human culture in which different people and groupings have particular rank according to their 'stage of development' (Yuval-Davis 1997) and therefore local cultural norms are ignored. Relativists strongly reject this approach and claim that different civilizations have different cultures, which need to be understood and judged within their own terms. This view means, in planning terms, taking into consideration cultural norms including those that subordinate women and abuse their human rights. One of the major criticisms of both approaches is that neither of them take account of internal differentiation and differences in positioning (Yuval-Davis 1997).

What is perhaps more relevant to the discussion in this chapter is to approach the debate about cultures and planning from a power relations point of view, looking at planning as the field in which power relations have their spatial expressions, as elaborated in Chapter 1. This chapter deals with the extent to which such power relations in planning affect its outcomes so that planning as a modern project serves as a double control over minorities in general and women in particular.

The chapter begins by providing a brief background on women and civil and human rights in Israel, with particular attention to women in Bedouin and Ethiopian communities. This is followed by a theoretical discussion on human rights and cultural relativism and its expression in space, planning and development. Then, planning for the Bedouin and for Ethiopian Jews and its effects on both men and women are examined, analysing the impact of the 'western-universalist' planning approach on these communities. The chapter concludes with a summary of the dilemmas and suggestions for further research.

WOMEN AND CIVIL AND HUMAN RIGHTS IN ISRAEL

Israel is a Jewish state which by definition categorizes all of its non-Jewish citizens as minorities. Being a Jewish state in a region where spaces are strongly classified has the result that one of the priorities of planning is control (Yiftacheal 1995). The situation leaves no room for any 'bottom-up' planning initiation either by Jews or non-Jews.

As for human rights in Israel, the state sets up foundation laws, some of which, such as the right to equality, are included in Israel's 1948 Declaration of the Establishment. This law states that all individuals are equal, no matter what their race, colour, gender, religion or political opinions. In addition, Israel has also ratified several UN Declarations of Human Rights such as the UN Convention of the Elimination of All Forms of Discrimination against Women (CEDAW). But in spite of these laws and ratifications, the desired equality for women in Israel has not yet been achieved.

Women make up half of the population of Israel, yet they are still treated as a 'minority' group, in their lack of equal access to resources and political representation. In spite of the state's commitment to 'freedom, justice and peace', expressed in the Declaration of the Establishment of the State of Israel, it is now realized that the gender equality whose existence has been proclaimed in Israeli society (especially in women's participation in the kibbutz and the drafting of women into the military) is a myth. Universal kinds of gender discrimination appear in Israeli society, including women's underrepresentation in the Knesset and in governmental, political and financial decision-making bodies, segregation in the workplace, undervaluation of women's labour, feminization of poverty and the abuse of women's bodies and their subordination through domestic violence, rape, sexual harassment and pornography (Shalev 1995). In recent years, the agenda for the advancement of gender equality can be gleaned from the subjects of recent legislation. In 1991 a Domestic Violence Act was introduced. In 1992 the Income Tax Ordinance was amended, substituting an optional gender-neutral taxation model for the old male breadwinner, head of household model; a social welfare act for single-parent families was passed; and employers were required to continue payment of certain employment benefits during the period of statutory maternity leave. The year 1993 saw the precedent-setting enactment of affirmative action legislation with regard to the appointment of members of boards of directors in government-owned corporations concerned with health rights, and a parental right to be absent from work in the case of a child's sickness was recognized (Shalev 1995).

But these advances are not sufficient to establish equality because religion in Israel dictates women's lives to a large extent. Jewish, Muslim and Christian women are still subordinated mostly on the basis of family laws, laws of marriage and divorce, which are still governed by religious courts in Israel. Both Jewish and Muslim religious laws perceive women as symbols of collective identity and as carriers of the 'burden of representation' (Yuval-Davis 1997). Because of patriarchal perceptions of women's fragile nature, their modesty and honour must be guarded more fiercely than men's. The result is that, according to Jewish and Islamic religions, women cannot be equal, laws of family subordinate them and their place is considered the private sphere of domesticity.

Muslim women in Israel sometimes face stronger restrictions because of the Shari'a family laws, which aim to preserve women's honour and modesty by strict laws of marriage and divorce. This can create deeply unhappy arranged marriages against women's will, and makes divorce almost impossible. Since Muslim sexuality is actually territorial (Mernissi 1987) rules of modesty which restrict women's movement in public and allocate space to each sex are very clear-cut and well kept, especially among the Bedouin. Muslim women's 'burden of representation' in Israel is sometimes fatal. Women have been murdered by their male relatives because of behaviour that brought 'shame' on their families and community. Each murder raises yet again a public discussion between the relativists and the universalists about what should be the official state policy and what the role of the police should be in preventing such crimes (see Rabinowitz 1995). In addition, cases of female circumcision still occur among the Bedouin in the Negev; these are not discussed in public and no statistical data exist on this matter.

As previously mentioned, Bedouin society's codes of modesty and honour are territorial. They dictate the boundaries of the 'forbidden' and the 'permitted', a matter of life or death in many cases. These norms or habits of seclusion, which are experienced in most of South Asia and in the Muslim world, contradict some basic human rights such as freedom of movement, the right to work and the right to political participation; sometimes they lead to extreme poverty and even death (Nussbaum 1995). These codes also contradict one of the foundation laws in Israel, the right to move in the country and the right to choose where to live (The Association for Civil Rights in Israel 1996). For Bedouin women, this right is abused twice. Firstly, as members of Bedouin society whose human rights are also violated twice, by massive land confiscation and by the lack of choice of one's own lifestyle (the Bedouin can only move to government towns or remain living in their traditional settlements in which they are then considered as illegal; see below). Secondly, Bedouin women's rights are abused by their own culture. Their freedom to move in the towns is more restricted than it used to be, traditionally, because the higher residential density in government-built towns increases the chances of unwanted meetings.

The Ethiopian community was brought *en masse* to Israel, as part of the 'national mission' of Israel as a Zionist state, to gather together diaspora Jews wherever possible from all over the world and bring them to the Jewish homeland. Ironically and tragically, one of the most problematic issues in the integration of the Ethiopian Jews in Israel has been the debate over their Judaism. The Ethiopians are officially recognized as Jews according to the decision of the Rabbinate, but despite this recognition, until the mid-1980s they had to undergo special conversion procedures in order to be registered on their identity cards as Jews (Corinaldi 1988). They are permitted to marry under Jewish law after undergoing ritual immersion (Rosen 1987). This policy harmed relations between the Ethiopians and Israeli society at the critical initial point of encounter and probably also had impact on gender cultural norms such as practices during menstruation, which had to be dramatically changed on coming to Israel. Most importantly, the Ethiopians point to the fact that they are black and that they have come from a Third-World country to a white westernized society as a significant cause of their difficulties.

Because control is often an important priority in spatial planning in Israel, it is most noticeable in development policies and planning schemes for non-Jewish groups, that is, Palestinian Arabs and the Bedouin. But to a large extent it is also noticeable in development programmes and plans for non-western Jews, i.e., the Oriental Jews in the 1950s and the Ethiopian Jews in the 1980s. In both cases a westernized universalist approach was adopted, which ignored the special cultural needs of these groups. Indeed, from its establishment in 1948, planning in Israel in many ways represents a western universalist attitude towards Jewish people as part of national universalist ideology of creating one western-oriented Jewish-Israeli culture. Towards Muslim and Christians, official planning policies were of control and partition, resulting in massive land confiscation and lack of financial public investments in Arab settlements.

WHY BOTHER? PLANNING FOR MUSLIM BEDOUIN AND ETHIOPIAN JEWISH WOMEN IN ISRAEL[1]

'Forbidden' and 'permitted' spaces in resettlement planning for the Bedouin

Half of the 100,000 Bedouin living in the Negev were forced to move to seven towns planned by the Israeli Government, which believed this strategy was necessary in order to provide the Bedouin with modern services and infra-structure. The rest of the Bedouin population live in some 108 traditional, tribal and clan-based villages; they are illegal in the eyes of the state, which therefore does not provide them with basic services and infrastructure. The two types of habitations differ from one another in their spatial characteristics, especially in matters of residential density. The traditional village is dispersed according to landownership patterns, that is, each tribe or clan resides on its own land with enough distance from other tribes to assure privacy, maintain the modesty of women and avoid unwanted meetings (for detailed analysis of traditional settlements, see Fenster 1993)

The planning of the towns, which took place from the early 1970s up to mid-1980s was a 'top-down' process, carried out by government-appointed Jewish planners with only a minor, 'lip service' participation of the Bedouin in the planning process. As a result the planning 'product' reflected the interests of the government rather than the needs of the population. This is mainly expressed in the selection of an urban type of habitation which served the government's aim, to concentrate the Bedouin in only a few large-sized settlements in order to minimize the allocation of land for Bedouin habitation. This decision was completely con-trary to the will of the Bedouin, who wished to 'legitimize' their rural traditional small-scale settlements. In addition to its political nature, the choice of towns rather than rural settlements expresses a western universalist approach which prioritizes western measures of efficiency even against the cultural needs and desires of the target population. The planners were aware of the problematic aspects of the 'product' – the town, which hardly met the needs of the Bedouin for tribal segregation and tried to meet those needs by translating cultural segrega-tion into spatial segregation. Each nuclear family lives in a separate house, with extended families grouped into an alley or cul-de-sac; a clan or a tribe are assigned to a neighbourhood, and a number of tribes live in a town (Kaplan and Amit 1979). By establishing this pattern, the planners tried to follow the tradition of segregation of tribes, to preserve the privacy of each social unit and minimize intertribal friction.

At the same time, the planners tried to plan a town which would be compact enough to facilitate proper provision of services. Such a contradictory task required compromise: on the one hand the traditional need for privacy within an expansive space, on the other the practical requirement of relative restriction in order to minimize costs. In other words, the planners tried to create a mixture, a universalist-relativist plan. This compromise was to be achieved in the planning of the neighbourhoods. Neighbourhoods are divided according to tribal affili-ation, and located on a distinctive topographical feature such as a hill, and at a large distance from each other. In each neighbourhood, each extended family receives a block of plots from the authorities. Plots are planned around cul-de-sacs and there is no through pathway from one neighbourhood to another, in order that there should be no unwanted interactions. At this scale of planning, it might be argued that the planners tried to combine a 'relativist' plan for neighbourhoods in a western universalist town structure. In reality, no matter how hard the planners tried to keep the distance between the neighbourhoods, after a while all neighbourhoods expanded and became fairly close to each other, so that the threats of abuse of cultural codes of honour increased.

This outcome and the greater housing density in the towns, (compared with their traditional settlements) show that the actual choice of urban habitation was wrong. It created a situation in which the transition to towns reduced the sizes of the 'permitted' spaces significantly. In the new towns most housing plots are located near to each other and are not greater than one dunam in size (0.1 hectare). Increasing residential density in towns, in comparison to the dispersed way of life in traditional settlements, drastically changed the boundaries between the 'forbidden' and the 'permitted', because of the greater fear that codes of honour would be violated. The effects of increased density have frequently been alluded to by Bedouin men. In a survey carried out among some 350 Bedouin men, more than 80 per cent of the respondents claimed that both plot size and distance between plots were small.[2] The fact that residential density reaches two to three housing units per dunam means that Bedouin women usually live with parents-in-law or with the husband's family; and such living arrangements increase the feeling of confinement. Housing density also restricts the degree of mobility in public space, because women are not free to move where men from outside their family congregate. Women who were interviewed in four focus groups expressed feelings of suffocation in this situation.[3] Not only do families live in close proximity to one another, but extended families very often live in the same house (or different floors of the same house), whereas in traditional settlements each nuclear family would have a separate domestic unit. In traditional villages, the 'public' area does not threaten codes of honour and modesty as it does in the towns because the chances of unwanted meetings are less than in the more densely built-up areas.

The result of the so-called 'modernized relativist' plan was that the confined neighbourhoods were the only 'permitted' spaces open to Bedouin women outside their homes. Most women can move freely only within their neighbourhood. They do most of their daily shopping for bread and dairy products within it, while clothing and shoes are purchased in the nearby cities. (City centres are 'permitted' areas because they allow a sense of anonymity.) The higher density within the neighbourhoods and their greater proximity have limited women's ability to enjoy the use of newly introduced services. Women do not feel comfortable in the public areas, such as the town centre. Moreover, parents do not always send their daughters to co-educational high schools for fear of unwanted meetings with males from other tribes, and the rapid spread of rumours about such meetings.

For most of the women, the remainder of the town is a forbidden space. Indeed, most women in the focus groups did not know any neighbourhoods other than their own. Bedouin women do not know or use services in other neighbourhoods, and most, aside from a few women in the largest of the new towns, have no social relationships in other neighbourhoods.

An indication of the ineptitude of 'universalist' modernized planning in creating appropriate public spaces for the resettled Bedouin is clearly seen in the fact that a large modern park-like area was built in one of the towns, away from the residential areas and located near the entrance road to town where everybody could watch the people in the park. For precisely this reason, nobody visited these areas. Assiza's story illustrates the fact that the complexity of culturally constructed spaces in Bedouin society creates forbidden spaces for women:

> One afternoon, my kids were bored and being naughty, as there is no playground in the neighbourhood, so I thought why not take them to the park that we have at the entrance to the town. You know, it is so beautiful, full of grass and flowers . . . very pleasant. After all why do we need this if we can't use it? So I went for the first time with my kids and after maybe ten or fifteen minutes my husband came hurrying to take me back to the neighbourhood. He was angry, and told me: 'Don't you know it is forbidden?' Apparently, his brother drove along the road and saw me, and went straight to my husband to tell him.

In Bedouin towns, public spaces such as parks have other functions than in western 'universalist' modern towns. They cannot be used for recreation, because of the increased risk of unwanted meetings. The net result of the resettlement is that Bedouin women's permitted space has shrunk, and their spatial mobility ends in their own neigbourhood. The next step in their spatial hierarchy are other, Jewish, cities in the region, which they visit occasionally because there the chance of unwanted meetings is smaller. The westernized relativist planning approach has worsened women's situation in relation to their right to move outside their home by providing planning solutions which are inappropriate for the population.

The increased restriction of movement in the newly-planned settlements abuses Bedouin women's right to work. Only 5 per cent of Bedouin women are employed (Ministry of Housing 1995), a low rate when compared to 44 per cent of Israeli Jewish women and 15 per cent of Israeli Palestinian-Arab women. The reason for this low rate is primarily cultural. After marriage, and especially after the birth of the first child, women stop working because social and cultural restrictions cut sharply 'permitted' spaces. Only a small percentage of women, those with high school education and beyond, work outside their homes. The few who work tend to be married to men who have had twelve years of education themselves, and are more open to modern lifestyles and tolerant of greater mobility for women. The desire to work is not always for survival only but for self-fulfilment, as illustrated in the following story of Halima:

> When I studied in high school I was so unique, not many girls studied at high school in my times. I dreamt that I would go to university to study mathematics, but my husband's parents did not allow me. Even to work at the high school laboratory they don't allow me. I was once unique and educated and now I am exactly like other uneducated women.

Restrictions on the right of work could perhaps be avoided if employment opportunities were constructed in the neighbourhoods themselves or at least in the towns. At present no employment opportunities exist in Bedouin towns and men have to commute to nearby cities for work.

The universalist approach may have some positive value for women, however; the move to towns has served as an 'accelerator of social change' and opened up male perceptions on their wives' rights to work. This conclusion emerges in men's discussions about their wives going out to work; with the move to towns, their views have become more flexible. Results of the survey show that most Bedouin men in towns (75 per cent) support the idea of developing employment opportunities for women in the settlement, whereas only 30 per cent of men in traditional settlements support such a proposal.

Spaces of purity and impurity in housing projects for the Ethiopian Jews

The 65,000 black Ethiopian Jews now living in Israel came from rural communities and arrived in two main waves: the first in 1984, and the second in 1991 in a dramatic rescue operation from civil war in Ethiopia, when some 19,000 people were brought to Israel in twenty-four hours. The Ethiopian Jews have undergone tremendous and dramatic changes in all aspects of their lives in the process of their integration as citizens into a white, western-oriented Israeli society. In Ethiopia they lived in an extended family structure, with the men working as farmers, blacksmiths and potters (Doleve-Gandelman 1990) and the women taking care of children, the household, and helping their husbands with agricultural work. They retained their own unique religious traditions. As immigrants, they face totally unfamiliar social, economic and religious norms and values. For the majority, the process of change has proved very difficult, especially with regard to gender relations.

Women make up on average 49 per cent of all Ethiopian immigrants, of which 46 per cent are aged 16 and above. About 28 per cent of the families in the first wave of immigrants (in 1984) were single-parent families, of which 83 per cent were matriarchal households, since there were many couples in which one partner stayed behind in Ethiopia or died on the journey to Israel. Long separation between couples led to infidelity, with no subsequent divorce, and this often created turmoil in the family structure.

For Ethiopian Jews, both in their homeland and in Israel, gender distinction is maintained by the tradition of the menstruation hut, the edge of whose territory is the boundary between pure and impure areas and between private and public space. In Ethiopia, women were separated during menstruation and after childbirth, and lived in the 'impurity hut' – the *yamargam gogo*, removed from their daily routine. This tradition derived from a strict reading of biblical law which says that a Jewish woman is to be considered impure for the seven days of her menstruation, and after giving birth. The women's relatives would bring food to the 'impurity hut' on special plates. Located close to the village, these huts were partially surrounded by a half circle of stones, which delineated the boundaries between the 'impure' and 'pure' spaces. Preferably located near a river, the huts also emphasized the Jewish identity of these women in villages where Jews and Christians lived together and were a means of ethnic distinction (Salamon 1993).

How did the planners in Israel take into account this habit in housing schemes for the Ethiopians? They did not! Although the customs of family purity upheld in Ethiopia were well known to the planners, they were not taken into consideration when immigrant housing designs were drawn up. Israel has no parallel for these particular meanings and use of spaces, so no comparable allocation of space has been allowed for in modern Israeli building schemes. Ethiopian women have lost a space which permitted them to remain closely in touch with their bodies (Doleve-Gandelman 1990). This rapid change has led to tension among family members, with the women especially talking of feeling guilty and ashamed. Many women have tried to improvise replacements for the hut, using balconies, hotel corridors or even a closet to separate themselves (Halper 1987; Westheimer and Kaplan 1992; Zehavi 1989). A small number seem to have found a solution to this problem in adapting the common Jewish rabbinical law about menstruation (*nida*), using the *mikve* (a pool-like ritual bath) in which to immerse themselves, to replace bodily purification in the running water of Ethiopia's rivers. Some women have tried to retain a modicum of separation in their home by allocating special rooms and building separate bathrooms for their menstrual period (Anteby 1996), but for most the loss of the ritual hut was a loss of a very important social institution. Anteby claims that Ethiopian women have lost their social role and identity by no longer being able to indicate with their bodies the boundaries between 'pure' and 'impure' spaces.

Interestingly, Westheimer and Kaplan argue that the fact that this custom has been preserved, even if in compromised form, while other customs are disappearing, should serve as a warning against any single interpretation of their meaning and function. Neither the feminist perspective, which attacks these customs as 'double subordination', nor the romantic approach, which defends them as a way of separating the woman from her daily routine, has really understood the rationale behind this custom. Precisely because of this lack of understanding, the preservation of this habit should be considered carefully. The ambivalence towards this habit might in some way be explained by the reactions of the Ethiopians themselves. Reactions within the Ethiopian community to the changes are mixed, with some women happy to be relieved of the menstruation hut, while others have nostalgic memories of this tradition. Perhaps the most significant and painful expression of loss is found in the feelings of some Ethiopians that Ethiopia is a 'clean' place and their present living environment is 'dirty' (Anteby 1996).

In the existing form of planning in Israel, there is no room for such different spatial needs. Thus what was 'public' in Ethiopia, and a part of each community, becomes relegated to the 'private' in Israel, with the individual having to replace the community in assuming responsibility for maintaining a separate 'pure' place.

In its policies and master plans, the state does not recognize the necessity to incorporate unique cultural or ethnic needs, and fails to provide an alternative for the customs and duties of the community, probably because planners assumed that these customs would quickly disappear with assimilation into the rest of society. Perhaps the planners also thought it was important to emphasize the similarities between the immigrant community and the society at large, rather than to highlight differences which are already expressed in different skin colours. Planners could, for example, have considered allocating separate apartments in each neighbourhood as a substitute for the menstruation hut in order to allow the immigrants to feel less threatened as they adjust to their new life.

In what sense does planning for the Ethiopians abuse human rights? It has forced the community to change some of its cultural values drastically and traumatically without involving it in the decision about whether these changes were wanted or whether they could be absorbed so rapidly. The effect of this universalist approach has mainly been to make relations between men and women in these immigrant households volatile. It has generated male violence, a phenomenon that was less common in Ethiopia.

The two case studies clearly present the ethical dilemma of whether to adopt a universalist or relativist viewpoint, and the negative effects of universalist plans on the situation of women of each group. The next section discusses this dilemma as it is analysed in the literature by way of suggesting alternatives.

HUMAN RIGHTS, CULTURAL RELATIVISM, UNIVERSALISM AND TRANSVERSALISM IN PLANNING AND DEVELOPMENT

In spatial planning there is no simple resolution to the relativist/universalist dilemma. Planning reflects what An-Na'Im (1995) calls 'the politics of culture'. Cultural norms and institutions are somewhat ambiguous and contestable; this permits various interpretations and practices and contradictory tendencies for both change and stabilization. The relationship between those with power and those maintaining cultural practice is often one of conflict (Yuval-Davis 1997). As a 'spatial mirror' of power relations in society, planning can become a very useful and influential way to do what Gasper (1996) calls 'mummifying' cultures, or to change them. This means that planning usually carried out by the powerful and dominant cultural majority can also serve to control space and control the well-being of both men and women, especially of cultural minorities. This approach takes planning as a process in which space is allocated; those who make the decisions are usually the powerful and dominant cultural majority.

As the two case studies show, women of cultural minorities are most likely to be made worse off by 'universalist' planning which does not take the cultural norms of women's honour or women's seclusion into consideration. The effects of planning in Israel on Bedouin and Ethiopian women emphasize the dominance of the western Askenazi universalist culture which asserts itself over, and sometimes controls, cultural minorities of both Jews (such as the Ethiopians) and non-Jews (such as the Bedouin). As illustrated in the previous section, planning not only affects women more than men but causes particularly negative consequences for women.

As planning can be such a powerful tool, the cultural position taken in the planning process is crucial in influencing the stability or power relationship and cultural dynamics in society, especially of gender relations. If planning internalizes cultural patriarchal codes, then women's subordination deriving from these codes will be perpetuated, whereas if plans are prepared with the goal of enhancing gender equality, it might be a means of change in gender relations. A cautious approach is necessary, however, when universalist planning is initiated, since it might worsen the women's situation, if the society, or the men, are not really ready for changing cultural norms, as the two case studies indicate was the case.

Amartya Sen (1995), Martha Nussbaum (1995) and Rhoda Howard (1995) are the leading advocates of universalism who take a clear view against cultural relativism. Each of them claims that human rights are considered by international law to be the right held equally by every individual by virtue of his or her humanity, and for no other reason. Universalists in general argue that relativists undermine the true meaning of human rights and leave the door open for severe abuse of individuals in the name of collective rights. It is 'cultural absolutism', says Howard (1995), when cultures are perceived as the sole source of the validity of a moral right or rule. Equality of women is one of the contested issues in the relativity of human rights. It is not accepted in many parts of the world.

In contrast, relativists hold the position that the notion of human rights is a western product, born out of liberal political philosophy. They claim that efforts to universalize human rights can be viewed as an act of imperialism or colonialism, rather than an act of liberalism. Cultures themselves can judge their own practices, argue the relativists. This position actually raises a very basic question as to whether human rights are collective or individual. Relativists argue that human rights are collective and that all cultures have some ideals of human rights, although these ideals may seem strange to westerners since they do not include norms, such as equality, which are the basis of the UN human rights framework. From this perspective, all cultures are equal, without any western superiority, and all cultures have their own human rights standards. Some scholars advocate the creation of an international human rights law but highlight the difficulties in such an agenda. An-Na'Im (1995) believes that it would be difficult to establish a principle of customary international law prohibiting all forms of gender discrimination, while Renteln (in Afshari 1994) suggests adopting only those universal cultural ideals shared by all cultures as a foundation for cross-cultural human rights. Both approaches seem to homogenize cultures, failing to acknowledge that there is no one 'culture' for everybody (Gasper 1996). 'Cultural needs' may be perceived differently by different members of cultures, especially with respect to those norms that differ in their effects on the various members of a culture. Gasper takes a rather practical view of incorporating 'cultural needs' into development. He discusses these issues within a wider framework of 'development ethics', which fragments cultures into their components and looks at the complexities of incorporating cultural norms in development and planning while emphasizing cultural difference. Gasper raises ethical questions regarding social change and suggests how to deal with it by looking at cultural components as sets of identities. McGregor (in Gasper 1996) calls it a 'matrix of identities' consisting of 'a range of overlapping possible cultural identities' and not simply 'a' culture. In this matrix, rows represent ethnic, religious, gender, ideological and urban/ rural identities and columns represent socio-economic or class identities. This is in line with the suggestion that social change is analysed as an outcome of the relationship between ethnic versus citizen identities, using three models (Fenster 1996): the assimilationist, the pluralist and the discrimination models. While these models suggest ways of identifying appropriate interventions in social change, focusing on societies' cultural needs, what perhaps needs to be more emphasized is the necessity of a dialogue within and between the different members of a culture and other cultures, especially between planners and beneficiaries if they do not share the same cultural values. This process is very much needed precisely because cultures are not just arbitrary collections of values, artefacts and modes of behaviours (Friedeman in Yuval-Davis 1997). Cultures are reflections of each individual's and each group's own standpoints with respect to change, the tendency for development and continuity on the one hand, and that of perpetual resistance to change on the other (An-Na'Im 1995; Yuval-Davis 1997). Because each individual or group, including the planners, have their own partial situated knowledge (Yuval-Davis), the necessity for participation and dialogue becomes crucial.

In the context of this chapter, participation and dialogue are especially relevant for planners who work with cultural traditions within which women's freedom to move in public is 'forbidden', restricted or seen as immodest. As the Bedouin case emphasizes, cultural norms such as purdah are used to designate what is 'appropriate' or 'respectable' behaviour for women (Chen 1995). These norms include the seclusion in their homes of the women for whom the planning is done, and the veiling of them in public. It is therefore necessary to have a dialogue of both men and women in the community, in order to make clear the cultural construction of space for them all and to plan in accordance to their needs and desires.

My personal view is that planning for cultural minorities from the human rights perspective is not an 'either/or' matter. It is neither the 'universalist' nor the 'relativist' approach that we should adopt, nor something in between, as becomes clear from the example of planning for Bedouin. As already mentioned above, what perhaps is needed is to establish participation, a dialogue between the planners involved and those who benefit from the outcomes of planning. The solution of a dialogue as a tool to sort out problematic situations in which religious and customary laws abuse human rights is proposed by various scholars (An-Na'Im

1995; Kandiyoti 1995; Yuval-Davis 1997). An-Na'Im sees the objective of what he calls 'cross-cultural dialogue' as bringing religious and customary laws into conformity with international human rights rather than extinguishing or trans- forming them. Changing popular beliefs and attitudes can be carried out, he argues, only by formal and informal education and by providing alternatives.

Alternative solutions or a plurality of solutions are perhaps key issues for the planning process. Changes can take place only if alternative plans are suggested, because it is only when variety of options exist that different needs can be real- ized. Later in this section I will mention some alternative planning solutions for both the Bedouin and Ethiopians. The type and nature of alternative plans depends on the extent to which room for manoeuvre exists in a community (Kandiyoti 1995). One of the ways to identify the potential for change is by applying transversals as an alternative to the universalism/relativism dichotomy (Yuval-Davis 1997). Transversalism is a form of a 'dialogue' in which all partici- pants bring with them awareness of their own roots and identities, but at the same time each tries to be flexible in mentally changing places with a man or woman who has a different role or identity in their society. Transversal dialogue allows differences to emerge within the shared interests of the group. Acknowledging difference necessitates what Kandiyoti calls pluralist policy in which every indi- vidual has the right to choose a lifestyle: 'the right of Muslim women in Bradford to veil and to attend religious schools, even with the support of public funds, does not deprive any other sections of the community of the possibility of making a different choice' (1995: 29). After all, the seclusion of women, which to western eyes is a kind of oppression, is seen by many Muslim women as a source of pride (Mernissi 1987) which they should be able to maintain side by side with other solutions. Because of the diversity of starting points, a great variety of choices in lifestyle is needed – to achieve the freedom of choice that is a fundamental human right; as Mahnaz Afkhami put it: 'rights provide individuals with choice and therefore the possibility of diversity' (1995: 5). The dialogue process in planning starts with identifying differences within the community and between planners and community members, and these differences are then worked out in a way to create choices. Differences exist not only on bases of gender, age, class and eth- nicity but also within these categories. Originally, the concept of transversal polit- ics dealt with differences among women, but this approach can be applied to differences between and within other groups.

A planning process which is an outcome of participation and a dialogue should produce choices, that is, a plurality of choices in a flexible form of planning. It means that plans can be 'relativist' and 'universalist' at the same time, so that, in the Bedouin case for example, there will be a large-scale park in a central location in Bedouin towns but also small-scale 'parks' or 'playgrounds' in each Bedouin neighbourhood. In this way, those women who wish to seclude themselves could use neighbourhood services and those who wish to, and can, challenge the restric- tion on their freedom to move could use central services. In the Ethiopian case, plurality of choices means, for example, the allocation of spaces for menstrual periods in Ethiopian neighbourhoods, such as menstrual flats in neighbourhoods where Ethiopians live, so that those women who wish to practise this habit separately from their home could have the option of doing so, while others who do not wish it will not be forced to do so.

The idea of a flexible-pluralist or 'multiple-choice' plan is that each component of the plan can be changed according to social and cultural changes in community members' attitudes towards cultural values which might result from transversal or cross-cultural dialogue. Here I take the viewpoint of Coomaraswamy (1995), Simon (in Gasper 1996), An-Na'Im (1995) and others who argue that for international human rights to be put into practice they have to go beyond the normative, textual essence and become a part of the legal culture of a given society and include efforts to change religious and customary laws. In this model only a dialogue with all members of the community and a review from within the culture itself provides the right perspective on its ability to absorb changes.

A useful tool with which to facilitate the translation of gender difference into planning and ensure a plurality of choices is the distinction between practical and strategic gender needs developed by Moser (1993). Strategic gender needs are those women identify because of their subordinate position to men in their society. Obviously, their content changes between different cultural contexts. They relate to gender divisions of labour, power and control and may include issues such as legal rights, domestic violence, and so on. These needs present the universalist viewpoint of, for example, planning from an egalitarian approach. In the Bedouin case, building a central park in Bedouin towns so that both men and women get accustomed to using public spaces without restrictions on women would exemplify such an approach. The decision to implement this use of space must be a result of a dialogue with the community members to find out whether these are indeed their needs, or at least the needs of some of the community members. Practical gender needs are the needs women identify in their socially accepted roles in society. These needs do not challenge the gender division of labour or women's subordinate position in society, although rising out of them. Practical gender needs are a response to immediate perceived necessities, identified within a specific cultural context, such as inadequacies in living conditions, for example in water provision, health care and employment. Practical gender needs reflect a 'relativist' viewpoint. Taking the Bedouin case as an example, practical gender needs require that the planning is focused on services such as playgrounds or clinics in each neighbourhood so that women and children can use them. This means accepting the cultural construction of space and women's lack of freedom of move. In a pluralist plan, however, the two options should exist, so that a change can take place towards a more egalitarian use of space for both men and women.

CONCLUSION

This chapter focuses on the dilemmas of incorporating cultural values in spatial 'top-down' planning, when these values abuse women's human rights. Two case studies of 'top-down' planning undertaken for the Bedouin and for Ethiopian Jews were presented. The analysis in the chapter shows how taking a 'universalist' view which ignores cultural norms of a community, increases the limitations on women's well-being even though these norms traditionally subordinate women. The deterioration in well-being occurs because of a lack of readiness of community members to accept changes. One conclusion that can be drawn from the chapter is therefore that planning for cultural minorities must include a dialogue among community members themselves and between them and planners, in order to identify the needs and wants of the different members of the community, their cultural perceptions of space and the plurality of choices that they need in order to ensure a smooth process of social change.

NOTES

1 An elaborated version of this section appears in: Fenster, T. (1998a) 'Space for gender: cultural roles of the forbidden and the permitted', *Society and Space*, and in Fenster, T. (1998b) 'Ethnicity, Citizenship and Gender: expressions in space and planning – Ethiopian immigrant women in Israel', *Gender, Space, Culture* 5(2), 177–89.
2 This survey was carried out in 1995 as part of the 'Development Plan for the Bedouin in the Negev' for the Ministry of Housing. The author wishes to thank the Ministry of Housing for permission to use this data.
3 Four focus groups were studied as part of the 'Development Plan for the Bedouin in the Negev'. Because of traditional reasons women could not be part of the survey and in order to hear their voices, these groups were created. The researcher met women from the same socio-economic background and they discussed various issues in an open-ended fashion.

REFERENCES

Afkhami, M. (1995) 'Introduction', in M. Afkhami (ed.) *Faith and Freedom*, London: I. B.Tauris Publishers.

Afshari, R. (1994) 'An essay on Islamic cultural relativism in the discourse of human rights', *Human Rights Quarterly* 16: 235–76.

An-Na'Im, A. (1995) 'State responsibility under international human rights law to change religious and customary laws', in R. Cook (ed.) *Human Rights of Women*, Philadelphia, PA: University of Pennsylvania Press.

Anteby, L.(1996) 'There's block in the house: negotiating female rituals of purity among the Ethiopian Jews in Israel', unpublished manuscript.

Association for Civil Rights in Israel (1996) *A Guide for the Rights and Obligations of Citizens in Israel* (in Hebrew) Tel Aviv: Association for Civil Rights in Israel.

Chen, M. (1995) 'A matter of survival: women's rights to employment in India and Bangaladesh' in M. Nussbaum and J. Glover (eds) *Women, Culture and Development*, Oxford: Clarendon Press.

Coomaraswamy, R. (1995) 'To bellow like a cow: women, ethnicity, and the discourse of rights', in J. Peters and A. Wolper (eds) *Women's Rights Human Rights*, New York: Routledge.

Corinaldi, M. (1988) *Ethiopian Jewry: Identity and Tradition*, (in Hebrew) Jerusalem: Rubin.

Doleve-Gandelman, T. (1990) 'Ethiopia as a lost imaginary space: The role of Ethiopian Jewish women in producing the ethnic identity of their immigrant groups', in F. MacCannell (ed.) *The Other Perspective in Gender and Culture*, New York: Columbia University Press.

Fenster, T. (1993) 'Settlement planning and participation under principles of pluralism', *Progress in Planning* 39: 167–242.

Fenster, T. (1996) 'Ethnicity and citizen identity in planning and development for minority groups', *Political Geography*, 15(5): 405–18.

Fenster, T. (1998a) 'Space for gender: cultural roles of the forbidden and the permitted', *Society and Space*.

Fenster, T. (1998b) 'Ethnicity, citizenship and gender: Ethiopian immigrant women in Israel', *Gender, Space and Culture* 5(2), 177–89.

Gasper, D. (1996), 'Culture and development ethics: needs, women's rights, and Western theories', *Development and Change* 27(4): 627–61.

Halper, J. (1987) 'The absorption of the Ethiopian immigrants: a return to the fifties', in M. Askenazi and A. Weingrod (eds) *Ethiopian Jews and Israel*, New Brunswick, NJ: Transaction Books.

Howard, R. (1993) 'Cultural absolutism and the nostalgia for community', *Human Rights Quarterly* 15(2): 315–38.

Kandiyoti, D. (1995) 'Reflections on the politics of gender in Muslim societies: from Nairobi to Beijing', in M. Afkhami (ed.) *Faith and Freedom*, London: I. B. Tauris Publishers.

Kaplan, Y. and Amit, A. (1979) *Master Plan for the Bedouin*, (in Hebrew) Tel Aviv.

Mernissi, F. (1987) *Beyond the Veil*, Bloomington, IN: Indiana University Press.

Ministry of Housing (1995) *Development Plan for the Bedouin in the Negev* (in Hebrew), Jerusalem.

Moser, C. O. N. (1993) *Gender Planning and Development: Theory, Practice and Training*, London and New York: Routledge.

Nussbaum, M. (1995) ' Human capabilities, female human beings', in M. Nussbaum and J. Glover (eds) *Women, Culture and Development*, Oxford: Clarendon Press.

Rabinowitz, D. (1995) 'Widening the path to rescue the mulatto women: an Israeli Forum' (in Hebrew) *Theory and Criticism* 7: 5–19.

Rosen, H. (1987) 'The Ethiopian Jews in Israel' (in Hebrew), in Beit Hatefutzot, *Beit Hatefutzot-Beita Israel*, Tel Aviv: Beit Hatefutzot.

Salamon, H. in (1993) 'Beta Israel' and their Christian neighbors in Ethiopia: analysis of central perceptions at different levels of cultural articulation' (in Hebrew), unpublished Ph.D. Thesis, Jerusalem: Hebrew University.

Shalev, C. (1995) 'Women in Israel: fighting tradition', in J. Peters and A. Wolper (eds) *Women's Rights Human Rights*, New York: Routledge.

Sen, A. (1995) 'Gender inequality and theories of justice', in M. Nussbaum and J. Glover (eds) *Women, Culture and Development*, Oxford: Clarendon Press.

Sibley, D. (1995) *Geographies of Exclusion*, London: Routledge.

Westheimer, R. and Kaplan, S. (1992) *Surviving Salvation*, New York: New York University Press.

Yiftacheal, O. (1995) 'The dark side of postmodernism: planning as control of an ethnic minority', in S. Watson and K. Gibson (eds) *Postmodern Spaces*, Oxford: Blackwell Publishers.

Yuval-Davis, N. (1997) *Gender and Nation*, London: Sage.

Zehavi, A. (1989) 'The integration of the Ethiopian immigrant women in the employment sector' (in Hebrew), in Employment Agency, *The Absorption of the Ethiopian Jews in Occupation*, Jerusalem: Ministry of Welfare.

4

INTERSECTING CLAIMS: POSSIBILITIES FOR PLANNING IN CANADA'S MULTICULTURAL CITIES

Marcia Wallace and Beth Moore Milroy

INTRODUCTION

This chapter addresses the challenges for planning in Canada's cities, which are becoming more and more consciously multicultural. Our point of view is that there are two main avenues that planners might take in meeting these challenges. One is to present diversity as an exception that needs accommodating over and above a 'normal' set of demands covered by the planning function. This is the current convention and is supported by societal structures, including the planning framework. Another option, however, is to recognise cultural diversity as already part of the basic make-up of cities, even though it is not often recognised in policies or institutionally based practices. Ideally, new structures could propose how variables such as gender, ethnicity and class could be treated together as intersecting one another; and how practice could be an exchange that produces something new rather than another opportunity for the dominant group to train subordinate groups in its ways. It is unclear to us which road will predominate in the future. In this chapter we describe how treating diversity as an exception is currently reinforced by broad societal structures and by planning frameworks, and offer some thoughts as to the implications of such a trend.

INTERSECTING CLAIMS: THE CHALLENGES FOR CURRENT POLICY

Neither the literature of planning nor recent experiences in planning practice make it obvious how one *ought* to deal with diversity in planning. Certainly the goals ought to be equity and fairness. But how to achieve these? Since the end of the Second World War, when urban and regional planning became widespread, there has been an increasing tendency around the world to define rights in terms of basic, universal rights attributable to all individuals, and to use the application of these as the means by which human rights are guaranteed for minorities (Kymlicka 1995: 2–6). However, concerns have emerged about the extent to which universal human rights can safeguard the rights of minorities, and whether supplementary protection needs to be available as well. These debates are the context in which planning is shaped. Planners find themselves in a new, evidently multicultural context in which questions are being raised about concepts that have long been associated with the movement for universal human rights, and to which they adhere, including official blindness to ethnic and gender differences.

In this regard, Canada has employed two concurrent approaches, which have until recently enjoyed wide support. One approach is to guarantee equal treatment of individuals before the law, including looking past ethnicity and gender; and the other is to recognise that equity can necessitate different treatment of individuals or groups, making the design and pursuit of ameliorating policies proper roles for governments. In line with the latter approach, initiatives have helped to improve the status of women in society by means such as pay equity legislation and policies directed at curbing violence against women. In the case of ethnic minorities, a multiculturalism policy was introduced in 1971 celebrating language and cultural diversity as a means of cultural preservation. The Multi-cultural Act, which followed in 1988, has served to legitimise claims by minority groups upon public policy-making and public funds.

However, a set of conditions currently exists amidst Canada's multicultural diversity which is rendering this web of approaches obsolete. One change in conditions is recent immigration policy, which is attracting high percentages of people with cultural backgrounds that are very different from the Euro-Canadian majority. In the past, immigrants arrived with customs not so socially distant from those already practised in Canada. While absorption was frequently described as difficult, the multiculturalism policy seemed a reasonable tool of integration by which Canadians invited immigrants to foster the customs that were important to them in order to be comfortable in their new home. As the face of immigration has markedly changed since the late 1970s, those with very different cultural, social, economic and legal traditions arrive now with much more complex needs, and there is resistance by the receiving population to some of the traditions. This is particularly the case for traditions that seem racist and sexist from the Canadian point of view; some of them have eventually been like a mirror in which Canadian society can see how its own traditions are exclusionary.

A second shift in conditions is that the multiculturalism policy is being widely challenged by ethnic minority groups and individuals themselves. Some argue that it is forcing differences upon immigrants when they prefer to meld into Canadian culture; others take the opposite view that multiculturalism is attempting to assimilate immigrants while purporting instead to recognise differences. Further, it is claimed that by paying attention to culture in the classic ways most often promoted by the policy, aspects such as gender and class are sidelined. This ignores the reality of the intersections of gender, ethnicity and class, and the considerable evidence gathered that shows ethnic minority status combined with gender creates even greater chances of oppression than either gender or minority status alone (Bottomley *et al.* 1991; Breitbart and Pader 1995; Gabriel 1996; Ng 1993; Stasiulis 1990; Young 1990; and others). The intersections are so complete in fact, that separate policies emanating from different government ministries do not serve as an adequate corrective. Indeed to recognise the multiple aspects of diversity is well beyond what current policy was designed to do. In the legal context Iyer explains that

> staffs of Canadian human rights commissions who investigate and prepare cases for adjudication, treat race discrimination and sex discrimination as separate and simple categories. Consequently, they regard the complainant (and her complaint) as relevant only to the extent that she falls within a particular category. This approach follows inevitably from current legal understandings of discrimination: sex discrimination is something different from age discrimination, and so on. There is virtually no consideration of the complex interactions of race, sex and the various other grounds of discrimination that are so much a part of the lived experience (as opposed to the legal analysis) of discrimination.
>
> (1997: 252)

This is why 'racial-minority women disappear from their own cases', as Iyer shows. A similar erasure operates in the planning field. This is the context in which planning practitioners, as well as other Canadians, draw their notions of what is legally appropriate regarding the treatment of diversity.

A third change in conditions pressuring the status quo is the reality that funding for policy accommodations is rapidly declining as the federal and provincial governments work their way through austere deficit-cutting agendas. This means that the interplay between law and policy, which had been oiled by extensive government funding, had managed to respond to many of the claims for justice by minorities. Now it operates less effectively to mask any structural problems. With fewer, less financially generous equity-seeking policies, individuals and groups will have to turn to other fora in which to make their claims.

The need for this discussion to occur at this time in Canada is only reinforced by the fact that national minorities, Aboriginal and Québécois, are seeking self-government or special status. In combination with the claims of ethnic minorities, these pressures are catapulting debate about group rights to the top of the national agenda.[1]

As Canadian cities become home to a visibly diverse population because of changes in immigration, a phenomenon described in detail in the next section, the ability to ensure equity in the face of real and potential racism and sexism is a task that national policy-making cannot in itself accomplish. Immigrants enter Canadian society at the level of localities, and so the effects of national immigration policy and human rights legislation must also be addressed at the local level. One tool at that level is urban planning, as will be explored later in this chapter.

A PORTRAIT OF CANADA'S MULTICULTURAL REALITY

In Canada, the cultural mix of local communities has historically been structured by federal immigration policy. Although Canada has long been a major receiving-country for the world's immigrants, it has never functioned with an open border. Policies exist which delineate, among other things, who shall enter, which countries of origin shall be considered, and what economic status among prospective immigrants is deemed desirable. In fact, during the first half of this century, potential immigrants from many countries were excluded, as some countries were considered more favourably by policy-makers than others. Chinese immigration, for example, was specifically restricted with the Chinese Act (1923) until 1946. Policy has also shifted with cycles in the Canadian economy, and has been reflected in priorities over entrance requirements alternating between Canada's economic benefit and immigrant family reunification.

In the latter part of this century, substantial changes to the Immigration Act have highlighted the effect of immigration policy upon the character of Canadian communities. Spurred by amendments to the regulations of the Immigration Act in 1962 and 1967,[2] and eventually the introduction of a new Immigration Act itself in 1975, immigrants from Third-World countries previously labelled 'restricted' or 'non-preferred' have replaced the once-European majority among Canada's immigrants. These immigrants introduced a spectrum of racial and ethnic diversity that had been previously unknown in Canadian urban communities. With these substantial policy changes which some consider to have removed racial barriers to immigration in Canada (Hawkins 1988: 11, 342), and others see as change in rhetoric to improve Canada's image abroad (Satzewich 1989: 85–90; Troper 1993: 266), immigration became an increasingly visible phenomenon in Canadian urban communities.

In this section, we illustrate the extent of Canadian immigration, emphasising the changes in ethnic and cultural composition since the mid-1970s. Specifically, we attempt to portray the multicultural reality that exists in Canada's urban centres and provide a context for understanding the complex intersections of ethnicity, gender and class that are the focus of this chapter.

A survey of the multicultural landscape

A relatively stable proportion of the population for several decades, immigrants[3] represented 16.1 per cent of the total population in 1991, or 4.3 million according to the national Census (Badets and Chui 1994: 5). Geographically, immigrants are mainly clustered in four of Canada's ten provinces (Ontario, British Columbia, Québec, and Alberta), and most reside within urban centres.

Urban areas hold obvious attraction for immigrants, typically offering a source of employment, community ties and a wide range of services. More than one-half (57 per cent) of Canada's immigrants, according to 1991 census data, live in the largest Canadian cities – the census metropolitan areas (CMAs)[4] of Toronto,

Montréal and Vancouver. This can be contrasted with the fact that just over one-quarter of the Canadian-born population live in those same CMAs (Badets and Chui 1994: 10). While these urban areas continue to be attractive to immigrants, settlement patterns are changing. In the Toronto CMA, the largest immigrant-receiving area in Canada, many newcomers are bypassing traditional arrival points in the urban core in the City of Toronto, in favour of the suburban cities within the surrounding regional municipalities (Vincent 1995a).

With the federal immigration policy changes of the 1970s, immigration into Canadian urban centres has undergone a significant shift in the source countries. Historically, Canadian immigration has come from European countries, but this has been steadily replaced by immigration from Asia and the Middle East (Figure 4.1). In fact, Asian and Middle Eastern immigrants account for six of the ten most frequently reported countries of origin (Badets and Chui 1994: 13). The Toronto CMA reflects this general pattern in Canadian immigration, and it is readily seen on city streets. Projections suggest as much as 45 per cent of the Toronto population will be made up of a combination of visible minorities by the year 2001 (Vincent 1995b).

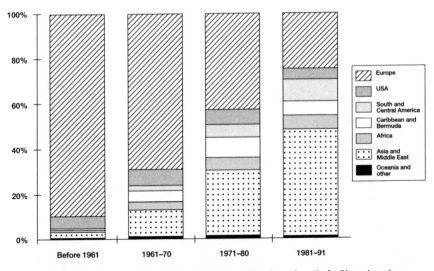

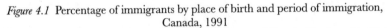

Figure 4.1 Percentage of immigrants by place of birth and period of immigration, Canada, 1991

Sources: Statistics Canada, *Immigration and Citizenship*; 1991 Census of Canada, Catalogue No. 93–316, Table 6

Intersections of ethnicity, gender and class

While multicultural diversity is a visible and significant change for Canada's urban centres, it is important to acknowledge that one's experience and identity are not mediated by ethnicity or culture alone. Within a multiculturally diverse city, the experiences and relationships of its residents may be structured by gender, class, age, religion, ability or other such categories, in addition to, separately from, or in combination with ethnicity. As this chapter focuses on the intersections of ethnicity, gender and class, it is relevant to describe Canadian society where these three categories intersect.

Although historically more men than women have immigrated to Canada, the ratio of male to female immigrants today is roughly the same as it is for the Canadian-born population; 96.3 men per 100 women and 97.9 men per 100 women respectively (Badets and Chui 1994: 35). According to the 1991 Census there are, however, considerable variations in the male/female ratio among immigrants when differentiated by country of birth. Figure 4.2 highlights this point. For example, there are more female than male immigrants to Canada from the USA, the Caribbean and Oceania. Men outnumber women, however, among immigrants from Africa and the Middle East.

Given the high priority placed upon the economic contribution of potential immigrants in Canada's immigration policy, it is significant to examine the distribution of immigrants and Canadian-born men and women in the labour force.

As Figure 4.3 indicates, immigrant men are more likely than Canadian-born men to be employed in professional, managerial and administrative occupations,

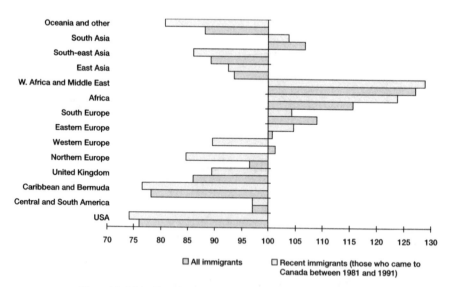

Figure 4.2 Male/female ratio* by place of birth, Canada, 1991
*The Male/Female Ratio is the number of males per 100 females.
Sources: Figures from Jane Badets and Tina W. L. Chui, *Canada's Changing Immigrant Population*, Focus on Canada Series (Ottawa: Statistics Canada and Scarborough: Prentice-Hall Canada Inc., 1994), 36. Compiled from 1991 Census of Canada, unpublished data

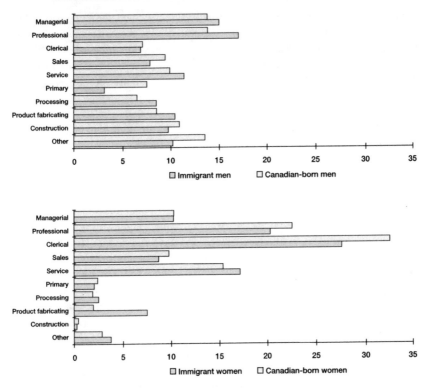

Figure 4.3 Percentage distribution of major occupation groups for Canadian-born and immigrant men and women aged 15 years and over, Canada, 1991
Sources: Chart reproduced from Jane Badets and Tina W. L. Chui, *Canada's Changing Immigrant Population*, Focus on Canada Series. (Ottawa: Statistics Canada and Scarborough: Prentice-Hall Canada Inc., 1994); 56, 58. Compiled from 1991 Census of Canada, unpublished data

as well as service, product fabricating and processing occupations. Immigrant women are more likely than Canadian-born women to be employed in service, product fabricating and processing occupations. Compared to their male counterparts, immigrant women are less dispersed among occupations, with a significant proportion employed in the clerical and professional occupations. This parallels the situation for Canadian-born women.

Although these occupational statistics offer some description of the inter-sections of gender, ethnicity and class in Canadian society, there are major omis-sions. They do not include all the work done in households and communities, both paid and unpaid, that exists beyond the formally recognised labour force. This work is especially relevant to a discussion such as this because it is where women, ethnic minorities, and lower-income people find themselves working in dis-proportionate numbers.

Perhaps the most invisible of this work outside the formally recognised labour force is the unpaid work done by women in households. According to the Statistics Canada General Social Survey of 1992, women still do two-thirds of all housework in Canada. Young couples tend to share the work more evenly, as do those in common law rather than married relationships. Women are spending 11 per cent less time now on unpaid work than they were 30 years ago, while their participation in the labour force has nearly doubled over the same period (Jackson 1996: 26–7; McDaniel 1994: 2, 64). At this time there do not appear to be data to show differences among ethnic groups regarding unpaid work.

Another example of work not included in employment figures is paid domestic service. The domestic service category is an illustration of the fact that in Canadian households work is not only an issue of gender, but also one of class and citizenship. Increasing participation by women in the formal labour force has boosted the demand for domestic workers. They are sought for housekeeping and childcare tasks which Canadian-born women tend to move away from when they can (Arat-Koc 1990). Some of the demand is filled by immigrant women doing house-cleaning, and being paid in cash on a household-to-individual basis. Other types of demand have been filled through Canada's Live-in Caregiver Program, where workers enter the country on a two-year temporary work visa rather than as formal immigrants. This program aims to attract foreign workers for live-in, paid domestic labour, and has brought workers from the Caribbean and the Philippines in particular over the last several years. Such foreign domestic workers, without citizenship or immigrant status in Canada, are 'neither wife nor worker', squeezed into a non-status limbo between the so-called private and the public spheres of Canadian society (Arat-Koc 1990: 86). Critics of the program argue that it has created an indentured servant class of immigrants: women without citizenship rights (Daenzer 1993: 135). Indeed, if these women stop working as domestics, thereby breaking the conditions of their work visa, they can be deported (Kim 1990: 179–80, also cited in Iyer 1997: 250).

The domestic worker program is a powerful example of how gender, ethnicity and class intersect, illustrating the complex ways in which Canadian immigration and multicultural diversity are interconnected in our cities. In the next section, we explore the human rights framework in Canada and its ability to protect the minority interests that are present amidst such diversity.

CANADA'S HUMAN RIGHTS FRAMEWORK

In Canada, human rights protection exists within a two-tiered legal system. At one level, anti-discrimination legislation is administered by quasi-judicial Human Rights Commissions in each of the ten provinces as well as at the federal level. The federal Human Rights Act (1978) protects human rights with respect to areas of federal jurisdiction (e.g. immigration, prisons), while provincial human rights codes operate within what is constitutionally considered provincial jurisdiction (e.g. municipalities, housing, education).

The second tier of human rights protection in Canada is the formal court system, where appeals can be made under the Canadian Charter of Rights and Freedoms (1982). Under the Charter, human rights are protected as constitutional rights, with the weight of the protection rooted in section 15 of the Charter which enshrines the right of equality under the law regardless of race, national or ethnic origin, colour, religion, sex, age or mental/physical disability.[5] Specifically relating to Canada's multicultural character is section 27, which states 'the Charter shall be interpreted in a manner consistent with the preservation and enhancement of the multicultural heritage of Canadians'. As the Charter goes no further in explaining this statement, however, this section is open to wide judicial interpretation.[6]

Within this legal framework, minority groups have claimed that their human rights have been infringed in a number of ways. For example, some important decisions have been made using the provincial level Human Rights Commissions (Andrew *et al.* 1994). In the province of Québec, a women's lobbying organisation working with housing activists and the Québec Human Rights Commission have defined harassment in housing, and some landlords have been convicted of discriminating against prospective tenants because they were single parents or poor. In Ontario, zoning by-laws were struck down that had restricted the number of people who could live in a dwelling who were unrelated by either blood or marriage. In 1990, claimants successfully argued before the Ontario Human Rights Commission that a condominium corporation discriminates on the basis of family status if it decides that dwellings in its multi-dwelling building may only be occupied by adults. The availability of housing for women and their children was at the core of the case.

Human rights protection has also been asked in Canada to recognise minorities that form around points where gender, ethnicity and class intersect in Canadian society. A 1993 Canadian Charter of Rights and Freedoms case in the province of Nova Scotia concluded that 'residents of public housing, the majority of whom are women, are a disadvantaged group protected by the Charter of Rights' (Hulchanski 1993). This decision makes it possible to challenge any municipality in Canada that attempts to restrict the amount of social housing within its jurisdiction. Elsewhere, in response to violence and abuse in the home against immigrant women who are brought to Canada within the family and fiancée sponsorship programs, human rights' and women's advocates have attempted to challenge the inequities they see imbedded in the framework of federal immigration policy (Kallen 1995: 117–18). Human rights have also formed the basis for challenges against language training programs open to those immigrants who intend to enter the labour force and have insufficient language skills (ibid.). Although the program does not directly discriminate on the basis of sex, immigrant women occupying jobs at the bottom of the labour market such as hotel maids or seamstresses are ineligible for such training programs because these tasks do not require the use of English or French.[7]

Equality but not necessarily equity

Because categories of difference like gender or ethnicity are not experienced in isolation, equality and fairness within a human rights framework are not easily ensured. Minorities, whether on the basis of gender, ethnicity, class or the like, cannot automatically access the protection a legal human rights framework provides, despite the apparent equity in application. For example, while everyone is formally protected from abuse and discrimination in employment, for immigrant women working alone in private workplaces as domestics, access to such protection may be prevented by other factors. The discrimination and exploitation many of these women undergo, including threats of deportation, low wages and poor working conditions (Boyd 1987), and fears about their immigration status or harsh consequences by their employers, can be tangible barriers for these women.

Other barriers, such as poor command of the language, lack of knowledge of the system, or fears of authority can also inhibit access to human rights protection for some people. For example, Property Standards by-laws are designed to ensure that housing is safe, yet despite the existence of formal complaint processes, many immigrants do not feel comfortable addressing violations in housing standards through the human rights framework. In a study of substandard housing in Kitchener, Ontario it was shown that when contending with unsanitary or potentially dangerous living conditions, immigrant men and women often choose to move when they get an opportunity rather than use human rights channels to improve their situation (Race Relations Committee 1992).

The challenge appears to be in the application of human rights protection. While most would readily agree that the human rights of all Canadians must be protected in an equal manner, there is less agreement as to how to insure that the legal framework will be equitable. It does not appear that a human rights framework that assumes universality can cope with the intersections of gender, ethnicity and class in Canadian society. While language for multiple diversities is worked out, it is unclear whether stronger minority rights, a reshaping of individual rights or some combination is the best direction to take. On the basis of the understanding that society and its public institutions reflect the culture of the majority, it can be argued that something more specific than human rights is required to ensure equal treatment in practice. These differentiated minority rights would not undermine principles of universality, but would offer justice between minority and majority groups (Kymlicka 1995: 47). Drawing a distinction between minority and majority groups can be useful in building an argument for rights protection, yet in reality differences stemming from gender, ethnicity or class make unpacking claims for equity a messy proposition.

In the next section, we describe the planning framework in one part of Canada. We note that it is applied in a largely universal and neutral fashion, similar to the application of the human rights framework. As will become obvious, planners are no further along in identifying means to address a multiculturally diverse public where ethnicity, gender and class intersect than those interested specifically in human rights.

THE PLANNING FRAMEWORK AND AN ASSUMPTION OF NEUTRALITY

While the principles of human rights and multiculturalism are established at the national level, their implications emerge concretely at the local level in homes, workplaces, civic organisations, and communities. In Canada, communities are planned from sub-national levels; a formal planning system at the national level does not exist.

We argue here that local planners shy away from recognising group differentiation occurring along ethnic, gender or cultural lines. If by necessity they do engage with this kind of social group differentiation, they treat it as an exceptional circumstance needing a unique remedy brought in from outside standard planning practice. By contrast, aggregates or associations representing economic classes and activities such as property owners, tenants and developers have long been ordinary and unexceptional participants who are defined as stakeholders in standard planning processes.[8] This proclivity for diversity-by-exception is explained both by where the planning function sits within the overall socio-legal matrix, and by the sources of the notions and theories that planners use about how communities ought, normatively, to be shaped. This point should become clearer as the chapter proceeds.

In Canada, specific powers have been assigned either to the federal government or else to the ten provincial governments.[9] Powers related to land use planning fall into provincial jurisdiction. This contrasts with many countries where planning is defined and determined nationally. Thus in Canada there are ten planning systems. This regional specificity afforded by provincial-level planning statutes is valuable in a country whose population is spread over such a gigantic and varied land mass – far too large for anyone to appreciate in detail and hence to plan for.

Canadian provinces delegate some of their power over land use planning to their municipalities, counties and regions. A tiered, or nested, concept of planning is used. When a local level plan is created, such as for a municipality, or lower tier government, its designers must pay heed to the plans and policies of the next level up, that is, of a region or upper tier government. In turn, regional plans must not contravene provincial guidelines or stated interests. The arrangement is such that municipal plans are intended to fit within metropolitan or regional plans, and these in turn to fit within provincial policies.

In general, planning in Canada has a strong land use orientation; areas such as health, education, social services, environmental quality and economy are dealt with separately from land use via other provincial legislation and local level institutional structures.

The case of Ontario planning

In order to look in more detail at the key features of local planning that are relevant to this discussion it is necessary to focus on just one of the ten provincial systems. We have selected Ontario. With about 11 million people, it has the largest population of Canada's ten provinces. Relevant to this chapter, however, is that Ontario receives just over half of Canada's immigrants, and its largest city, Toronto, has been a leader in the Canadian planning field for several decades.

Planning legislation in Ontario has evolved since its inception early in the twentieth century in terms of the content of plans, relationships between levels of government, and relationships between councils and citizens. However, its fundamental hierarchical structure, its regulatory and its land use orientation, have remained. This is also characteristic of the planning systems of other provinces. Where the provinces differ from one another is in the specific content of the various planning legislations, the dispute resolution mechanisms provided, and the agencies involved in implementation. Because the provinces' planning laws share basic features and also sit within a common two-tiered national human rights system, it is not surprising that, broadly speaking, all local planning functions in comparable ways. This means that gender and multiculturalism issues will have similar reverberations in all provinces across Canada.

Under Ontario's Planning Act (Ontario 1996), legislation establishes that official plans must be produced, and that they must contain certain elements. For instance, official plans must include: goals, objectives, policies and a description of how these will be achieved; how land is to be subdivided; how citizens are to be advised about the adoption and amending of local plans; and the structures and procedures available to contest planning decisions. When being developed or amended, local plans must have regard to a wide range of designated interests of the provincial government such as for housing, wetlands, and food-producing lands. This referral requirement is a primary means by which contradictions between government levels and agencies are managed.

Local municipalities (which may be cities, towns, villages or townships) design plans and send them to a higher government level for approval. They are approved if they fit within the general legislation and upper tier plans and guidelines. Once plans are approved, local jurisdictions are in charge of implementing them. Implementation requires sets of tools, including zoning by-laws, which are also described in the Planning Act. Reviews of official plans are called for every five years.

The planning legislation prescribes that citizens be informed about various changes being contemplated (e.g. new building or subdividing of properties) and in particular when an official plan is to be amended. The minimum conditions of notice are spelled out, including the number of days' notice, whether by mail, by posting notices in newspapers, making documents available at the council's offices, or holding a public meeting. Recent revisions to the Planning Act introduced the option for municipalities to set out alternative procedures in their official plans for informing and obtaining the views of the public. Citizens can comment on planning decisions by participating in meetings sponsored by local councils or by writing to local councils before decisions are taken. If redress is sought, one can appeal to a tribunal called the Ontario Municipal Board, which is devoted to resolving land use disputes; or, if appropriate, appeal to the Ontario Human Rights Commission which administers the provincial Human Rights Code, as noted previously in this chapter.

Analysis of the language used in Ontario's planning legislation reveals that people are described in the Planning Act as generic and undifferentiated. References to 'persons' or 'the public' are found throughout. The only collectivities mentioned that fall between the scale of the individual and the entire 'public' are 'public bodies', and these refer to a very limited range of civic collectivities such as school boards and public utilities. The parties affected by planning decisions are therefore assumed to be the individual, or the public as a whole, described as an aggregate. Further, the language of the Act conveys the assumption that any party is equally capable of enunciating concerns about planning processes and decisions and having them heeded. Patently this is not the case. In fact, there is considerable evidence to the contrary demonstrating who is disadvantaged and why by undifferentiated planning processes.[10] In contrast to planning *legislation*, which continued to be phrased within conventional universal rights traditions in each of several recent revisions to the Act, the planning *profession* and *practice* show occasional modest excursions into conceptualising diversity in planning contexts.

In 1994 the national *professional* body, the Canadian Institute of Planners, adopted a Statement of Values, as well as a new Code of Professional Conduct. The Statement pays attention to diversity while the Code does not. In the Statement there are injunctions such as 'to value the natural and cultural environment', 'to respect diversity', and to foster 'meaningful public participation by all individuals and groups and seek to articulate the needs of those whose interests have not been represented' (Canadian Institute of Planners 1994). The Code of Professional Conduct, however, uses the same generic 'public' as found in the legislation: for example, 'practice in a manner that represents the needs, values and aspirations of the public' (ibid.). An important distinction between the two is that the Code can be enforced by the Institute, with contraventions attracting sanctions, whereas the Statement of Values only proposes guidelines for action.

A 1996 case illustrates the squeeze in which planners find themselves in *practice* between the 'generic-leaning' planning legislation and enforceable Code of Professional Conduct on the one hand, and the 'diversity-leaning', unenforceable, professional Statement of Values and emerging human rights decisions on the other. The case concerns Old Order Amish who have lived in farming communities in southern Ontario for nearly two hundred years (*Mornington [Township] v. Kuepfer* 1996).[11] Their cultural and religious practices preclude using electrical power and owning automobiles. They get around by horse and buggy. This necessitates stabling horses wherever they live. In one of the townships where they make up a quarter of the population, there is a 1982 by-law prohibiting the stabling of horses in a residential zone. The township is mainly rural. Members of this group occasionally live away from the farms in hamlets where residential zones have been designated. A complaint generated an official order to two families living in a hamlet to cease stabling their horses in their barns – barns which had been built fifty years earlier.

The plaintiff in the case asserted that the by-law was based solely on secular planning criteria, which included zoning based on use, not based on people, resulting in the same treatment applying to everyone. The court disagreed; it said: 'Land use practices are made by human beings and are made with human beings in mind as well as land resources' (Canadian Institute of Planners, para. 77). The court found:

> that the policies of both municipal and provincial forms of government are exerting an indirect form of control over the lifestyle of the Old Order Amish whose religious beliefs play a very major role in their lives and thus limit alternative courses of conduct for the Amish otherwise available to the population at large.
>
> (para. 87)

Drawing on earlier Charter of Rights and Freedoms cases, it was noted that 'identical treatment may frequently produce serious inequality' (para. 81); and 'the interests of true equality may require differentiation in treatment' (para. 82). The court added that, 'the Charter accommodates positive discrimination to allow for the special needs of people . . . special religious and ethnic needs, such as the Old Order Amish, in a way that supersedes any zoning by-law' (para. 83). In the end, the by-law was found to be in violation of the Charter. The decision was not appealed.

This decision should give the planning profession pause. The case shows that strict adherence to the Code of Professional Conduct is not enough; the public is not a generic entity. The Statement of Values is needed to inform the application of the Code. Indeed, as was raised in this case, in neighbouring Wellesley Township, where a sizeable proportion of the population is also Old Order Amish, a solution had been found. Noting that about 20 per cent of its population was limited to horse-and-buggy transportation, and keeping in mind that one of the township's basic principles was to be mindful of the needs of all its citizens (para. 60), the local planner said that the township had created 'mixed-use residential/ agricultural clusters'. In effect the court concluded that the Wellesley Township approach more accurately reflected the intention of the Charter.

Another well-documented example of social group recognition within planning *practice* is associated with the safe city initiative, which originated in the late 1980s in Toronto and was subsequently taken up in several other Canadian cities. It resulted from women taking action against violence against women in public places. Being physically assaulted because of one's gender is an undisputed basis for defining social group affinity. Work began by focusing public attention on sexual assaults, and by participating in the development of Toronto's official plan, which was completed in 1991. The outcome was that safety must be considered during the plan's implementation. To that end, planning and development guidelines have been produced, and regular workshops are available for planners, police, architects, transit officials, and community activists to learn about safe city planning (Wekerle and Whitzman 1994; Whitzman 1994). Needless to say, this change only came about because of substantial and sustained pressure on the political and bureaucratic systems. However, it is a success story because it now appears widely accepted that planning to increase safety from violence against women is a 'legitimate' demand that can be made upon the planning system by a social group.

That same Toronto official plan contained a number of other important features (City of Toronto 1993). It was expressly given a social equity orientation[12] – which in itself is rare in planning. In spelling out what this meant, specific urban rights were identified. Besides the right to safety, which has been well used by the safe city activists, another is freedom from discrimination. One of the strategies in the plan that responds to this right is zoning to reduce land use segregation, which has long been high on women's agendas. Another strategy is enhanced pedestrianisation. It seeks to get people out on the streets and on to public transit. It is a planning-based action directed toward keeping different social groups in informal contact with each other on the theory that collective use of public spaces makes the differences between one another less strange. If less strange, then less threatening; if less threatening, then less likely to generate discrimination. It is one way of employing land use to serve social goals. Fortunately the 1993 plan for the Metropolitan Toronto region also recognised how important cross-cultural understanding is to the well-being of the region at this time in its history, and it identified mixed-use development, pedestrianisation and transit as tools to assist that objective. Consequently, the City of Toronto plan nests neatly within the regional plan as required by the provincial government. These are modest achievements resulting from major efforts on the part of a few people to nudge the institutional Leviathan on to another course.

FUTURE POSSIBILITIES FOR PLANNING AND DIVERSITY

Like most people, planners are only beginning to recognise that diversity is not only basic to the make-up of this country but also that this fact requires conscious attention if practices are to be up-dated. Currently, difference, whether associated with ability, income, gender or ethnicity, is treated as an exceptional circumstance to be accommodated within norm-focused planning processes (Qadeer 1994). Planners who want to engage in diversity initiatives find few conceptual tools and documented approaches, and fewer still when it comes to working with the intersections of various dimensions of difference, which should be the preferred goal. There is great need for case studies of practice in this field.

Meanwhile, two options appear to be emerging. The first is to incorporate diversity into traditional planning practice on a case-by-case basis, as noted in the previous section. Indeed, the discussion was intended to illustrate some explanations for why planners treat diversity in this way: the blueprint exists in planning legislation, in past practice, and in the Professional Code of Conduct. Operationally, this option improves access for minority groups to the planning process through measures such as increased outreach and accommodating specific demands made by pressure groups. Repeated practice begins to incorporate the diversity into regular practice without tampering with widely held ideals of universal equality. Qadeer (1994) describes the experience at the City of Scarborough[13] regarding processing commercial standards and site requirements for so-called 'Chinese shopping centres' between 1985 and 1990. Over time 'a mediatory and accommodating stance that is sensitive to cultural elements of these shopping centres' evolved. All the while the 'formal policies remain "blind" to ethnic and cultural characteristics of a proposal' (1994: 192). Elsewhere Qadeer (1997) illustrates this approach to diversity in the housing sector.

The second option is to take diversity as the point of departure. This means to see the city as already constructed by the existing diversity whether or not this construction has resulted from purposefulness or inattention, or whether it has been 'bad' or 'good'. The blueprint for this kind of planning is undeveloped. It requires more than a moral obligation towards inclusivity, and instead demands an understanding that diversity is fundamental to who we already are. One feature it needs is for people to come to the planning process 'as themselves', rather than as a generic person, or only as a representative of a particular, human rights-defined social group. While selecting one axis of difference, the 'whole person' is made to disappear, as Iyer (1997) has illustrated. The whole person is wanted at the planning table. Municipal responsibilities must continue to include input from those affected by a planning initiative, and to find ways to hear and work with collections of diverse people. While this approach is hardly more than a glint in the eye, there are hints regarding procedure in the recent official plan processes noted for Toronto and its region. They show that planners and others need to propose goals for public landscapes in which positive expressions of already-existing social diversity can be played out.

Which approach will prevail in the long run is a matter of speculation. In the short run, gains are welcome using either because the issue of diversity then becomes part of the public debate. While planners can provide legs for good ideas, they are far more likely to do so when spurred on by activists capable of linking lived urban experience to progressive theories from fields such as human rights, cultural studies, feminism and design.

NOTES

1 For the purposes of this chapter we follow Kymlicka's distinction between national and ethnic minorities (1995: 10–26). A national minority is, in Kymlicka's words:

> a historical community, more or less institutionally complete, occupying a given territory or homeland, sharing a distinct language and culture. A 'nation' in this

sociological sense is closely related to the idea of a 'people' or a 'culture' – indeed those terms are often defined in terms of each other. A country which contains more than one nation is, therefore, not a nation-state but a multination state, and the smaller cultures form 'national minorities'.

(p. 11)

Canada has both Québécois and Aboriginal national minorities within an English majority and thus it is a multination state in Kymlicka's terms. In addition, there are immigrant groups in Canada which have different claims on the country in that they are not 'nations' and they typically assert 'their ethnic particularity . . . within the public institutions' of the majority society, whether English or French in Canada (p. 15). This makes Canada also a polyethnic state. Our focus in this chapter is on ethnic rather than national minorities.

Kymlicka applies 'culture' and 'multiculturalism' to national and ethnic characteristics. He does not include women, gays and lesbians, or the disabled specifically, though he intends that his theory of cultural rights of minorities would be 'compatible with the just demands of disadvantaged social groups' (p. 19). Using his framework, we try to draw attention in this chapter to the tools available to recognise ethnic minorities within planning contexts, and ethnic minority women in particular.

2 In 1962 the regulation changes involved the removal of 'preferential' treatment (categories of 'preferred' and 'not preferred' countries), replacing it with a focus on economic qualifications (education, training, skills). In 1967 the regulations were again amended, this time to formalise the economic focus with the introduction of the 'points system' in place today (Hawkins 1988: 11, 342).

3 In Canada, an 'immigrant' is defined as a person who has come to Canada to be a permanent resident, but has not yet become a Canadian citizen. With such status, people can move freely across Canada's borders, and take part in programs of universal access, including medical care and legal assistance. Canadian citizenship is granted automatically to those born in Canada, and it can be achieved by immigrants and refugees through naturalisation.

4 CMA is a designation used by Statistics Canada to represent the main labour market area of an urbanised core with a population of 100,000 or more. The Toronto CMA, for example, is a large geographic area that includes the City of Toronto, the cities within Metro Toronto and neighbouring towns, villages and townships.

5 As with all Charter rights, this protection is weakened by the subjective interpretation of section 1, which states these rights and freedoms can be guaranteed subject to 'such reasonable limits prescribed by law as can be demonstrably justified in a free and democratic society'. For a discussion of this point see: Kallen 1995: 260; Mandel 1994: 41.

6 Section 27 of the Charter has been criticised as nothing more than a 'motherhood statement', an interpretative provision as opposed to a substantive guarantee (Kallen 1990: 173).

7 Two Charter challenges have been raised on behalf of immigrant women against the language training program and the fiancée sponsorship program. They have argued that these federal government programs violate the equality rights of immigrant women under the Charter (s.15). *Canadian Human Rights Advocate*, March 1988, cited in Kallen 1995: 117–18.

8 The very helpful distinctions developed by Iris Marion Young (1990: 421–8) among social groups, associations and aggregates inform our terminology. Briefly, 'a social group is a collective of persons differentiated from at least one other group by cultural forms, practice, or way of life'; members have an affinity; and groups exist in relation to one another (p. 43). An aggregate is a classification of individuals according to an attribute, e.g. dwelling type or age. Associations are formally organised institutions such

71

as clubs, corporations and unions. Associations and aggregates conceive of the individual as 'prior to the collective' whereas social groups 'constitute individuals' through a sense of group identity – not wholly, but nonetheless in significant ways.

9 There are two territories, the Yukon and the Northwest Territories, which have less autonomy than provinces and which at this time are redefining their relationships with the federal government and the country as a whole as part of the settling of land claims with First Nations peoples. One outcome is that the Northwest Territories will be divided into two in 1999, making three Canadian territories. Nunavut, meaning 'our land' in the Inuktitut language of the Inuit, is the new name for the eastern portion. Nunavut occupies almost 25 per cent of the Canadian land mass, with a population of just 25,000.

10 To give just four Canadian examples of disenfranchisement: see Shkilnyk 1985 on an Aboriginal group; Helman 1987 on a low-income group; Piché 1988 on low-income women and girls. An American example on racial and ethnic minorities can be found in Hoch 1993.

11 Special thanks to Professor Edward Bennett for bringing this case to our attention.

12 The social equity orientation found in Toronto's official plan can be attributed to a dynamic and diverse group of interests being brought together as a Task Force, and their report: *Goals and Principles for a New Official Plan* (City of Toronto, 1990).

13 On 1 January 1998, Scarborough, five other cities and the Metropolitan government amalgamated to form the new City of Toronto.

REFERENCES

Andrew, C., Gurstein, P., Klodawsky, F., Milroy, B. M., McClain, J., Peake, L., Rose, D. and Wekerle, G. (1994) *Canadian Women and Cities*, Ottawa: International Relations Division, Canada Mortgage and Housing Corporation.

Arat-Koc, S. (1990) 'Importing housewives: non-citizen domestic workers and the crisis of the domestic sphere in Canada', in M. Luxton, H. Rosenberg and S. Arat-Koc (eds) *Through the Kitchen Window*, 2nd edn, Toronto: Garamond Press.

Badets, J. and Chui, T. W. L. (1994) *Canada's Changing Immigrant Population* (Focus on Canada Series), Ottawa: Statistics Canada; Scarborough: Prentice-Hall Canada (Catalogue no. 96–311E).

Bottomley, G., de Lepervanche, M. and Martin, J. (eds) (1991) *Intersexions: Gender/Class/Culture/Ethnicity*, New South Wales: Allen and Unwin.

Boyd, M. (1987) *Migrant Women in Canada: Profiles and Policies* (Immigrant Research Working Paper no. 2), Ottawa: Employment and Immigration Canada.

Breitbart, M. and Pader, E.-J. (1995) 'Establishing ground: representing gender and race in a mixed housing development', *Gender, Place and Culture* 2(1): 5–20.

Canadian Institute of Planners (1994) 'Statement of Values and Code of Professional Conduct', by-law amendment, Ottawa: Canadian Institute of Planners.

City of Toronto, The Task Force (1990) *Goals and Principles for a New Official Plan*.

City of Toronto, Planning and Development Department (1993) 'City of Toronto Official Plan Part I – Cityplan', approved by council, July 20, City of Toronto.

Daenzer, P. (1993) *Regulating Class Privilege: Immigrant Servants in Canada, 1940s–1990s*, Toronto: Canadian Scholars' Press.

Gabriel, C. (1996) 'One or the other? "Race", gender, and the limits of official multiculturalism', in J. Brodie (ed.) *Women and Canadian Public Policy*, Toronto: Harcourt, Brace.

Hawkins, F. (1988) *Canada and Immigration: Public Policy and Public Concern*, 2nd edn, Montreal/Kingston: McGill-Queen's University Press.

Helman, C. (1987) *The Milton-Park Affair: Canada's Largest Citizen-Developer Confrontation*, Montreal: Vehicule Press.

Hoch, C. (1993) 'Racism and planning', *Journal of the American Planning Association* 59(4): 451–60.

Hulchanski, D. (1993) 'And housing for all: opening the doors to inclusive community planning', *Plan Canada* (May): 19–23.

Iyer, N. (1997) 'Disappearing women: racial-minority women in human rights cases', in C. Andrew and S. Radpers (eds) *Women and the Canadian State – Les femmes et l' état Canadien*, Montreal/Kingston: McGill-Queen's University Press.

Jackson, C. (1996) 'Measuring and valuing households' unpaid work', *Canadian Social Trends*, Ottawa: Statistics Canada (Autumn) 25–9 (Catalogue no. 11–008-XPE).

Kallen, E. (1990) 'Multiculturalism: the not-so-impossible dream', in R. I. Cholewinski (ed.) *Human Rights in Canada: Into the 1990s and Beyond*, Ottawa: Human Rights Research and Education Centre, University of Ottawa.

Kallen, E. (1995) *Ethnicity and Human Rights in Canada*, 2nd edn, Toronto: Oxford University Press.

Kim, K. (1990) *Domestic Workers' Handbook*, Vancouver: Legal Services Society.

Kymlicka, W. (1995) *Multicultural Citizenship: A Liberal Theory of Minority Rights*, New York: Oxford University Press.

Mandel, M. (1994) *The Charter of Rights and the Legalization of Politics in Canada*, 2nd edn, Toronto: Thompson Educational Publishing.

McDaniel, S. A. (1994) 'Family and friends', *General Social Survey Analysis Series*, Ottawa: Statistics Canada 9 (August) (Catalogue no. 11–612E).

Mornington [Township] v. *Kuepfer* (1996) Ontario Court of Justice (Provincial Division), No. 1724.

Ng, R. (1993) 'Sexism, racism, Canadian nationalism', in H. Bannerji (ed.) *Returning the Gaze: Essays on Racism, Feminism and Politics*, Toronto: Sister Vision Press.

Piché, D. (1988) 'Interacting with the urban environment: two case studies of women's and female adolescents' leisure activities', in C. Andrew and B. M. Milroy (eds) *Life Spaces*, Vancouver: University of British Columbia Press.

Ontario (1996) *Planning Act*, Revised Statutes of Ontario. Office consolidation containing extracts from Bill 20. Royal Assent, April.

Qadeer, M. (1994) 'Urban planning and multiculturalism in Ontario, Canada', in H. Thomas and V. Krishnavayan (eds) *Race, Equality and Planning*, Aldershot: Avebury.

Qadeer, M. (1997) 'Pluralistic planning for multicultural cities', *Journal of the American Planning Association* 63(4): 481–94.

Race Relations Committee (1992) *Substandard Housing in Kitchener-Waterloo: A Focus on Ethnic Minorities*, Kitchener, Ontario: Race Relations Committee, Kitchener-Waterloo.

Satzewich, V. (1989) 'Racism and Canadian immigration policy: the government's view of Caribbean migration, 1962–1966', *Canadian Ethnic Studies* 21(1): 77–97.

Shkilnyk, A. (1985) *A Poison Stronger than Love*, New Haven, CT: Yale University Press.

Stasiulis, D. (1990) 'Theorizing connections: gender, race, ethnicity and class', in P. Li (ed.) *Race and Ethnic Relations in Canada*, Toronto: Oxford University Press.

Statistics Canada (1991) *Immigration and Citizenship*, 1991 Census of Canada (Catalogue no. 93–316).

Troper, H. (1993) 'Canada's immigration policy since 1945', *International Journal*, 48(2): 255–81.

Wekerle, G. and Whitzman, C. (1994) *Safe Cities*, New York: Van Nostrand, Reinholt.

Whitzman, C. (1994) 'In Toronto, planning is the best defence', *Planning*, 60 (1):10–11.

Young, I. M. (1990) *Justice and the Politics of Difference*, Princeton, NJ: Princeton University Press.

Vincent, I. (1995a) 'Chasing after the ethnic consumer', *Globe & Mail*, September 18.

Vincent, I. (1995b) 'Ethnic communities definitely on the rise', *Globe & Mail*, November 1.

5

THE GENDER INEQUALITIES OF PLANNING IN SINGAPORE

Gillian Davidson

INTRODUCTION

International definitions of human rights have been inextricably linked to the position of minority groups and most commonly outlined in terms of ethnicity (Evans 1990). When planning for multicultured societies, therefore, priority has been given to ensuring equality and freedoms for the various ethnic groups by placing the creation of a stable and racially-tolerant society high on the agenda of most nations in their attempts to industrialise and foster economic growth (Hettne 1996). In South-east Asia, this mix of ethnic identities and the increasing physical mobility of populations are perceived as forces that might destabilise the process and goals of economic development by generating animosity, discrimination and violence (Chiew 1990). Singapore is emblematic of a nation which has not only successfully combated the divisive and disruptive tendencies of a plural society, but simultaneously risen to become one of the most successful newly industrialised nations in the world and a favoured model for human investment strategy by both developing and industrialised nations (UNDP 1995). This success has been achieved by the very detailed attention to planning in all aspects of economic and social life by the ruling People's Action Party (PAP), which has retained power since independence in 1965. By building on the concepts of productivity, meritocracy and racial harmony and appealing to the notions of communitarianism and national survival, the PAP has implemented a series of economic and social planning strategies, such as export-oriented industrialisation and nation-wide public housing, in order to successfully marry ethnicity with development (Teo and Ooi 1996).

In the 1990s the PAP has focused this planning on cultural values and, in particular, on a national ideology for Singapore, reflecting fears of the erosion of Asian culture and, therefore, ethnic tolerance and economic dynamism. Definitions of the new Singaporean identity have been based on a core set of selective and traditional 'Asian values' from each of the main ethnic groups, promoting the family, community and nation above self. Women have been integral to these planning agendas with the government actively defining and redefining gendered roles and relations according to these specific economic, political and cultural needs (Doran 1996). Such detailed attention to planning for racial tolerance and economic success, however, has received criticism from the international community, which dubs this style of government as intrusive and authoritarian (*Financial Times* 1995; Rodan 1993); critics highlight the denial of the freedom of equal and full rights for all (*Straits Times* 22 August 1994). This has also led to criticisms in Singapore of the expanding gap between liberty and wealth (Heng 1994) and the consequent downgrading of human rights. Crucially, the importance of such freedoms and rights is rejected in Singapore (and many other non-western states) as a form of cultural imperialism and supremacy, defending their paternalistic actions in this case as a new form of democracy and political vocabulary evolving in Asia, which subordinates the individualism of the West for the social obligation and morality of the collective (Devan and Heng 1994). Senior Minister Lee Kuan Yew says, 'The fundamental difference between Western concepts of society and government and the East Asian concepts is that Eastern societies believe that the individual exists in the context of his family. He is not pristine and separate' (quoted in Ching 1994: 38). Such notions of family and collectives mean that until recently, the differential effects of such pro-active state planning on a number of different groups in Singapore have been ignored and only slight attention has been given to the impact on the rights and equalities of women (for an exception see Davidson 1996; Davidson and Drakakis-Smith 1997, on poverty).

Feminist analysis has criticised preoccupations with ethnicity, cultural values and neutral planning because of the resulting neglect and disregard for women, which pushes them to the margins of planning (Sullivan 1995). Proponents of women's rights have, therefore, rejected the naturalness of culture and its ascribed roles and categories by attempting to problematise the very definitions upon which culture and its values are based and sought to disaggregate the collective to reveal internal inequalities and violations. They identify these cultural definitions and premises as main violators of human rights for women, arguing that they are used by both state and private actors as tools to justify and defend acts of inequality, discrimination and exclusion against women (Rao 1995) regardless of ethnicity, language, race or culture. By inculcating a distinct set of value systems and codes of behaviour, cultural definitions in turn determine gendered roles, relations and social and spatial practices in everyday lives (Bunting 1993; Kaufman and Lindquist 1995). In the case of Singapore, identity and culture and the definitions of women as mothers, workers and citizens have been consistently (re)constructed in line with current economic and planning demands. The experience of minority positions in the family, community and nation is, therefore, socially conditioned and integral to the hegemony of political power relations. However, few human rights advocates in general have sought to protect the interests of groups or individuals who are at odds with the control and regulations of state authorities and government actors, and especially so if the individuals or groups are women.

In this chapter I consider how the preoccupation of state planning with both economic growth and multiculturalism impacts on the human rights experiences of different groups of women in Singapore.[1] Whilst Singapore is an international success story, commentators have warned that progress and economic growth do not necessarily secure or ensure human rights, such as equality, inclusion and empowerment, for all. Many of the socio-economic changes and policies implemented by the PAP will have undoubtedly benefited women in Singapore, but others are more ambiguous since women are used as tools for economic or political advantage. Drawing on two examples from social planning by the state in Singapore which focus on the definition of what it is to be a 'woman' and a 'mother', this chapter aims to highlight how the construction and inculcation of cultural values, morals and gendered roles as tools for planning impact on the fragile nature of human rights and equality for women. I begin, however, with a brief background to planning, the state and cultural development in Singapore.

THE CULTURAL DEVELOPMENT OF SINGAPORE

Planning for success

Upon independence in 1965, the government of Singapore inherited a plural society of Chinese, Malay and Indian settlers,[2] groups who were ethnically diverse and distinct from one another. Furthermore there was a rapidly growing population, widespread poverty, unemployment and economic uncertainty. However, by the 1990s, Singapore had launched itself into the ranks of the middle-income economies as a leading exporting and industrialised nation (World Bank 1991). The social and economic statistics bear witness to this achievement with a GNP per capita of US \$10,450 in 1990, unemployment rates under 3 per cent and literacy rates at 91 per cent (all in 1993) and the successful eradication of the divisive tendencies of a multiracial society (Department of Statistics 1994; World Bank 1991). Success and development have also been shared by all groups, with impressive displays of democratic achievements uniting ethnicity, class and gender, such as the massive Housing and Development Board (HDB) estates.

Kong (1995: 449) has attributed many of these economic and social achievements to the 'shrewdness of the People's Action Party (PAP), not only in its economic strategies but also in its social engineering, including its conscious attempts to shape values and political cultures'. The PAP's implementation of detailed policies and interventionist planning was to achieve very clear and selective economic and social goals which would significantly change the structure of the economy as well as the organisation of society and the population. All policies have primarily been brought in to promote economic growth based on attracting foreign investment, doing this by combining a free market ideology and export-orientated industrialisation, together with controlled human resource management (Drakakis-Smith *et al.* 1993; Rigg 1991). This pro-active state management was necessary in order to extend and intrude into all aspects of daily life, both public and private, from investment strategies and national savings to filial care for the elderly and childbirth; these actions have been justified and defended by the PAP by referring to the rationale of pragmatism and realism in planning for a multiracial community (Soin 1996). However, the political and selective nature of planning in Singapore creates policies which are intrinsically discriminatory and, therefore, potentially exclusionary.

Women have been inextricably bound to these development agendas and planning processes (Wong 1981) and in particular used to strengthen state goals. This is most clearly demonstrated by the importance placed on women and the family as instruments of social change and by the development and promotion of women's various roles such as worker, mother and daughter at different stages in the development process. For example, in the early stages of Singapore's independent history in the 1960s and 1970s, the pursuit of labour-intensive, export-oriented industrialisation focused on increasing the numbers of women in the workplace whilst curbing the naturally high fertility rates. However, by the 1980s attempts were made to upgrade the economy to capital-intensive and hi-tech industries to compete with economic rivals with a skilled, quality workforce. In contrast to other developing countries and against usual demographic trends, this also included the stimulation of higher fertility rates and a return to the emphasis on women's roles as mothers to meet expanding job vacancies. Furthermore, women were promoted as carers in the family in attempts to cope with a rapidly ageing population. At this time the government had perceived an impingement of external influences on Singapore's open society resulting in the erosion of Asian cultural values and racial harmony. The PAP, therefore, set out to ensure the population's productiveness and national loyalty by focusing planning on the themes of appropriate cultural values and national identities and the place of the family.[3]

Constructing culture: the search for Singapore's national values

The use of culture, and in particular the promotion of a neutral, communal, Asian culture, has been integral to Singapore's economic and social planning for a multiethnic nation. This practice became particularly important in the late 1980s when the government noted a deterioration in values and morals and feared increasing threats from the perceived individualism of the West and disruptive tendencies of Islamic fundamentalism. Such a shift was viewed as a threat to national harmony, productivity and security, provoking the government, 'to immunise Singaporeans from the undesirable effects of alien influences and to bind them together as a nation' (Quah 1990: 1). Such 'undesirable effects' included the erosion of the work ethic, which would undermine the efficiency of the nation, and an increasing economic and social dependency on the state, which would burden the government with expanding social costs. Whilst implementing strategies of industrialisation and modernisation, therefore, the PAP simultaneously set out to preserve traditional and cultural values in Singapore in order to restore what they identified and defined as national interests. Relying on this conflation of the concerns of government and society, the PAP selected a particular set of cultural values upon which the construction of the Singaporean national identity could be based and to which the Chinese, Malays and Indians could be persuaded to direct their loyalties (Chun 1996; Willmott 1989). In 1989, President Wee Kim Wee outlined the government's vision and this chosen set of core Singaporean values,

> If we are not to lose our bearings, we should preserve the cultural heritage of each of our communities, and uphold certain common values which capture the essence of being a Singaporean. These core values include placing society above self, upholding family as the basic building block of society, resolving major issues through consensus instead of contention, and stressing racial and religious tolerance and harmony. We need to enshrine these fundamental ideas in a National Ideology. Such a formal statement will bond us together as Singaporeans, with our own distinct identity and destiny.
>
> (quoted in Quah 1990: 1–2)

Primarily, these cultural values were those of sharing and community; they were emphasised in order to unite the different groups, and were promoted through educational programmes and public campaigns. By advancing the social and psychological defence of the nation, Brown (1994) writes, the state created a national 'garrison mentality'. This is exemplified by the militaristic group spirit of campaign slogans, such as 'Team Singapore', 'Total Defence' and 'One People, One Nation'. Secondly, national identity was grounded in the importance of the family and filial responsibility as the building block of society. This approach was an attempt to preserve traditional Asian values but it also allowed the state to transfer the increasing burden of welfare, and especially an ageing society, on to civil society and primarily the family, in order to keep public spending low and to continue to attract foreign investment. Both of these assumptions have massive implications for women and their daily experiences, since they required specific constructions and re-constructions of women's identities, roles and relations. The focus on collectivity and sameness fails to recognise the difference and diversity of women's experiences in both their public and private lives and shows a disregard for the impact of planning on women (Tomasevski 1993), whilst the focus of cultural planning on the family fails to disaggregate households on the basis of gender (Moser 1993) and inevitably affects women both directly and indirectly because they continue to find themselves at the centre of family life (Singham 1992).

Recent women's rights and feminist commentaries have criticised this type of inculcation of cultural norms in planning and the use of women to strengthen state goals. They argue that this allows both the state and private actors to justify inequalities, discrimination and abuse against women and to deny women's rights the status of human rights. Some of the issues identified as gender-based abuses include the inequalities of opportunity, for example, in education, housing and employment; domestic violence and reproductive rights and the devaluation of childcare and domestic labour (see Peters and Wolper 1995). Rao (1995: 169) asserts that 'no social group has suffered greater violation of its human rights in the name of culture than women', and as a consequence they are the world's largest excluded and minority group (Karl 1995) regardless of ethnicity, race, class or culture. The importance of cultural definitions and premises is their appearance of naturalness and universalism and their ability to construct and institutionalise values and norms. Doran (1996: 153) asserts that 'identity in Singapore is not a natural-given, nor has it been historically stable. Rather it has been discursively constructed and reconstructed within particular historical contexts'. By the very nature of this artificiality, the creation of culture is subjective and judgemental; gendered roles are manufactured and preserved, as are relations and everyday practices. When implemented as part of a planning agenda this invoking of culture becomes an inherently political act (Moser 1993), where the ability to define and construct culture is dependent upon position, status and power relations in society. Rao (p. 173) says that 'Culture is a series of constantly contested and negotiated social practices whose meanings are influenced by the power and status of their interpreters and participants.' Massey (1994) suggests that such displays of power and control are exemplified in the exercise of 'boundary-drawing', a spatial act which is intrinsically socially and culturally qualified and potentially exclusionary. This is especially the situation when boundaries and margins are strongly framed and classified to create descriptive and discriminatory categories and spaces, such as the public and the private spheres, which in turn serve to conceal significant social inequalities and oppressions (Sibley 1992). The ability to contest and negotiate these practices and spaces, and therefore power relations, is bound to how culture and places assign categories and define the social construction of a 'woman'.

In Singapore the categorisation of dichotomous spaces and identities is particularly evident (Davidson 1996). Here, as in most other places, women are constructed in terms of the private sphere and spaces associated with the home, family and reproduction as opposed to the masculine productive and public sphere. Wong comments on how this perception transcends all ethnic groups:

> The traditional religiocultural systems of Confucianism, Hinduism and Islam prescribed a subordinate status to women within the household. Whether born Chinese, Indian or Malay, a woman was socialised from a young age both to play the role of wife, mother and daughter-in-law, and to lead a secluded life.
>
> (Wong 1981: 449)

It becomes clear that despite the dramatic changes in Singaporean society and economic advances, 'In an Asian society like Singapore, women are subjected to multiple demands at home. They are expected to be good mothers, filial daughters and obliging wives. These demands are compounded as more women go out to work' (Cheung 1991: 85). In the assigning of different roles and spaces to women and in their reinventing and remodelling of themselves to suit particular policies, there will be contradictions and anomalies! This sort of planning is not merito-cratic and fails to allow the true participation that is presented by the state. This failure will in turn compromise the quality of rights and freedoms for women.

The basic objectives of human rights and development paradigms have con-sistently been defined as 'enlarging people's choices' (UNDP 1995: 1) as expressed through an individual's ability to choose and participate in decision-making. In Singapore, it is this very diversity and choice that is denied by the powerful and restraining boundaries of social norms, spatial dichotomies and state planning, by means of which the family, and women's roles within it will be made to conform with the state's economic agenda. Without recognition of their multiple realities, women become silenced and invisible and, therefore, further removed from the achievement of equal status and rights. Grounding theory and planning practice in actual experiences and realities of women, therefore, has been regarded by many feminist commentators as essential in order to represent 'the voices, decisions, desires and acts of resistance of the unempowered' (Wolf 1990: 67), therefore allowing them to 'focus on less dramatic, unofficial, private, invisible spheres of social life' (Vaiou 1992: 258).

In the next section I focus on two specific state planning agendas in Singapore in order to investigate the impact of constructing and framing women's identities on their experiences of human rights and equal participation. These are examples of social planning which highlight women in their 'natural', yet constructed, family and maternal roles and show how state planning has intervened in different ways in the generally overlooked private, family sphere (Moser 1993). First, I look at population and fertility planning where women, in their roles as mothers, have been placed at the centre of planning and the national agenda, and second, at female single parents who in their one-parent family units represent to the Singa-porean authorities an increasing 'problem' and 'threat' and have been placed at the very margins of political planning and society.

THE INEQUALITIES OF PLANNING

Singapore's population policies

The use of fertility management in population planning is now a recognised component of social policy in many developing nations. However, reproduction has also been applied as a political tool where women are used as 'state pawns' to strengthen state goals (Kabeer 1995). Population control is one of the few areas of policy where women are not marginalised or pushed to the boundaries of plan-ning but find themselves as the central focus. Such attention implies positive benefits for women: expected increases in choices and control over all aspects of their lives and in particular their bodies. Feminists argue, however, that despite this focus on women, in reality women find that their role is reduced to that of mother and childbearer (Tomasevski 1993) which in turn compromises levels of equality, choice and participation. Rather than empowering, emancipating or even 'giving a voice' to women, population policies have rendered women invisible in aspects of their lives other than as mothers. Furthermore, such policies also fail to acknowledge that, not only do women experience motherhood differently and diversely but, that some never experience it at all (Tomasevski 1993). In human rights discourse, population policies have increasingly been considered violators of women's human rights, as policies act to deny women basic claims to their own bodies and fertility and therefore to control and power (Moser 1993). In contrast to the feminist literature, development theory presents Singapore as an example of successful socio-economic development and strong family planning and as a model for developing nations around the world (World Bank 1993).

In Singapore, women and their fertility have been used in a number of ways to achieve economic growth and racial tolerance in the state's 'nation-building project' (Drakakis-Smith *et al.* 1993). Population planning was undertaken as part of the PAP's drive to create a nation attractive to foreign investment. The result is that any changes in economic strategy, such as the move from labour-intensive to capital-intensive development, have been mirrored by equally extensive temporal and spatial modifications and turn-arounds in population policies. Women were not just integral to the labour force in terms of labour power but also in their roles within the home as mothers and childbearers since they were perceived as the source and nurturers of the nation's future labour supply. However, even this interpretation was not consistent, since particular aspects of mothering were emphasised as the requirements and demands of economic policy fluctuated.

Action was taken in the early stages of development to curtail fertility, a policy implemented in many developing economies to foster economic development. In the years between 1965 and 1984, population planning in Singapore was based on fertility reduction, through campaigns in family planning, sterilisation and legalised abortion using such slogans as 'Stop at Two' and 'Two is Enough'. The aim of this blanket policy was to discourage large families, which were popular in all ethnic communities whether Chinese, Malay or Indian, by encouraging small families through a series of state incentives and disincentives. For example, after a family's first two children, maternity leave was restricted, delivery fees were raised progressively with the number of children and priority access to school places was lost for the third and subsequent children (for more detail see Saw 1990). The result was a vast movement of women into the workplace. This campaign, coupled with rapid economic development, was immensely successful and resulted in total fertility rates dropping from 4.66 in 1965 to an unexpected low of 1.4 in 1986 (Teo and Ooi 1996). However, the success of these programmes eventually began to create problems for sustaining the population at replacement level; additionally, it coincided with the economic downturn of the mid-1980s. As a consequence, policies were instigated to move attention away from numbers in the workforce to the importance of skills and the quality of labour. This new direction in turn fuelled a turn-around in population planning and the reinvention of women's roles in order to increase fertility levels whilst at the same time

81

increasing the nation's skill-base. In a controversial statement, the government proclaimed that the philosophy of the new agenda was based on the principle that intelligence was inherited and in particular from the mother (Kong 1995). Attempts were therefore made to encourage the rising number of young single female graduates and other educated women to increase their fertility levels whilst discouraging the less educated, i.e. those with no O-Levels,[4] from having large families. Selective pro-natalist programmes included the Graduate Mother's Scheme which favoured higher marriage and fertility rates for tertiary-educated women; it offered, for example, priority access to school places for third and fourth children and increased tax relief worth up to S$10,000[5] for each of the first three children. Women with no O-Levels or earning under S$1,500 per month were subject to cash incentives for sterilisation and punitive delivery fees (Drakakis-Smith *et al.* 1993; Teo and Ooi 1996).

The third and current stage in population planning is a modified extension of this policy. The premiss of inherited intelligence still exists but barriers have been relaxed as initial attempts to increase replacement levels failed after the achievements of the 'Stop at Two' campaign. The need to increase fertility was perceived to be especially critical as the economy continued to grow whilst the workforce base began to shrink and the natural population base aged. Furthermore, with targets to phase out the reliance on foreign workers, the state had to reconsider its population agenda if escalating job vacancies were to continue to be filled. In 1987 the pro-natal 'New Population Policy' was introduced, aiming to promote higher fertility within most groups. However, with a slogan of 'Have three, or more if you can afford it', class and racial overtones first evident in the 1980s Graduate Mother's Scheme had obviously not been removed (Teo and Ooi 1996). Despite a slackening in eligibility levels, financial incentives such as tax rebates for third and fourth children continued to be reserved for more educated women, usually with at least three O-Levels. Meanwhile, the government maintained its need for women in the workforce in order to keep increasing vacancies filled. Again the target was more educated women, in line with the upgrading policies that included incentives such as subsidies for childcare open only to women who were working. For low-income families childcare costs, even with the subsidy, exceeded potential earnings, forcing mothers to remain in the home and out of productive employment, thus compounding problems of low income (Davidson 1996). These class differences and exclusions were also apparent in government schemes that were brought in to help needy families. For example, as part of the eligibility for the 'Small Families Improvement Scheme' (SFIS), applicants could not have more than two children and were required to sign legal documents pledging to stop after two births.

These different phases of population planning clearly demonstrate how women are used as tools of state planning in Singapore and are examples of discrimination, inequality, exclusion and denial of the freedom to choose. Furthermore, the focus of policy on the mother alone fails to acknowledge the role of both the father in family life and aspects of traditional culture such as the promotion of male children, still common in the Chinese community, and large families in the case of Islam. These types of Asian values, however, did not conform with the economic agenda of the state and therefore did not feature as part of Singapore's cultural values and national identity. Population planning has offered some women increasing freedoms and choices but has stripped others of fundamental and basic rights whilst failing to recognise the diversity of needs, desires and motivations. Tomasevski (1993) notes the special attention which has been paid to women through these policies, but only as childbearers in what have increasingly come to be recognised as 'breeding campaigns' and 'genetic engineering' programmes (Financial Times 1995). Women in Singapore have been denied the right to their own bodies and choices over how they are used in fertility and national planning.

However, it is clearly not gender alone which determines women's responses to planning or how they experience policy. Rather, gender is intertwined with other facets of development such as class and ethnicity. Kabeer (1995) writes that population policies should not be exclusive or selective in their benefits but should be concerned with the needs of everyone, including the poor. In the case of Singapore it is the very selective nature of planning and the assumptions of cultural values which have had the effect of sustaining poverty and vulnerability by excluding groups of women such as those who do not attain the education levels expected by current criteria. Women and their roles as mothers, therefore, are not constructed neutrally or impartially as is claimed, since some are ascribed more importance than others. One such group of women who are attributed lesser importance are single mothers; they are one of the most obvious examples of differentiation and exclusion in Singapore.

Constructing the family: the marginalisation of single mothers

In the promotion of a Singaporean national identity, government propaganda clearly outlined the importance of Asian cultural values and maintaining cohesive family ties:

> Today, families all over the world are exposed to value systems which undermine family life. There is a need to recognise and promote values which uphold the importance of family ties and thereby contribute to the collective good.
>
> (National Advisory Council on the Family and the Aged 1995: 4)

This is perceived to be integral to the economic and social survival of the nation, as Prime Minister Goh Chok Tong explained in his fourth National Day rally speech: 'if we lose our traditional values, our family strength and social cohesion, we will lose our vibrancy and decline' (*Straits Times* 22 August 1994). Media campaigns and family education programmes therefore have abundant appeals for 'positive values' and 'cohesive and healthy families', to emphasise the importance of marital relationships and filial piety.

The implementation of such precise and narrow definitions of the family can be identified as an example of Massey's 'boundary-drawing' because it neglects and deliberately ignores the differences and dynamism inherent in family life and the construction of households. Through the promotion of family life education, marriage enrichment programmes and state planning, the government has deliberately worked on the assumption of intact two-parent families, so actively discouraging single-parent households. Drawing on the experiences and per-ceived 'social ills' of the West, and in particular the USA and Britain, the Prime Minister made clear that he will not accept the proliferation of single parents, criticising the 'misguided government compassion towards single parents [which] had led to disastrous welfare policies in the US' (*Straits Times* 3 September 1994). Numbers of single parents in Singapore, however, remain low. There are an estimated 18,000 one-parent households[6] in total and in comparison to the US, only one out of a hundred babies is born illegitimate in Singapore compared to one in three in the US (*Straits Times* 29 July 1994). Divorce rates have also fallen in the 1990s, reflecting stricter laws and referral to counselling rather than a decline in applications. It is now well recorded that the vast majority of the heads of single-parent households are women and that their numbers are increasing. Furthermore, evidence world-wide reveals that most experience economic hardship, characterised by low income and wages, poor access to the labour market and insecurity and vulnerability (Kodras and Jones 1991; UNDP 1995; Varley 1996). In 1989, the Advisory Council on Family and Community Life in Singapore identified female-headed households as one of the main groups risking destitution. One report estimated that nearly half of single mothers have incomes below S$1,000 per month[7] (*Straits Times* 29 July 1994) whilst over 30 per cent have problems with rent arrears (Quek 1992). Success at improving conditions, however, is limited because of the barriers put up against the full and equal participation of single mothers in the workplace and the marginalising of them in planning and society.

State planning and laws are assertively pro-family as part of measures to reinforce the family and preserve selective traditional moral and family values. In turn this focus excludes and exempts single parents from several schemes since they are perceived to represent the deterioration of the family and therefore undermine national values and morals. Obstacles and barriers for single parents, however, are often not explicit but subtle and implicitly exclusionary. For example, single parents face obstacles such as the lack of provision of affordable childcare places, a necessity if the mother is to gain access to the workplace; only 200 low-cost childcare places are available for an estimated 5,850 children (not including the children of intact disadvantaged families who compete for the same places) (Quek 1992). Furthermore, as in the population policies, there is discrimination against the poorer households; to be eligible for childcare subsidy the mother must be working. Such restrictions on access and participation in the labour market can in turn lead to longer-term problems because non-workers and low-earners are unable to contribute to the compulsory savings scheme, the Central Provident

Fund (CPF), which is the only form of security for old age or source of payment to purchase a house in Singapore.

The pro-family stance also extends more explicitly into areas of social and welfare planning such as housing policy. It is used to encourage families to stay together by excluding single mothers from purchasing subsidised HDB housing. Furthermore, the Small Families Investment Scheme (SFIS), which aims to help low-income families through home ownership and education bursaries, is only open to 'low-income couples or widows'[8] (Ministry of Community Development 1996: 8), discriminating not only against single parents but between the different types of single-parent households. Also, if a family which was receiving benefits from the SFIS breaks up then penalties such as repayment of housing grants and interest are imposed and educational bursaries are discontinued. All grants and cash payments, however, are paid into the women's account as part of the SFIS scheme, making her liable for any repayment. Help and advice is available for single parents in the form of two voluntary agencies; however, between them they can only provide for 470 families (Goh 1991). These selective pro-family laws and planning agendas, and other examples such as the Graduate Mother's Scheme, reveal not only the intrinsically discriminatory and exclusive nature of planning, but also how unrelated benefits and assistance are to need, and how the 'deserving needy' in Singapore are defined.

This exclusion of single parents from mainstream planning and policy-making has been justified by Prime Minister Goh on the grounds that 'we must have this condition to prevent the proliferation of single mother families' (*Straits Times* 19 March 1993). Giving aid to single parents would be a display of acceptance and therefore condone the breakdown of Asian values and morals. What this exemplifies is Brown's (1994) 'garrison mentality', brought about by the fear of economic failure and the desire to succeed. The government's stringent criteria for eligibility for the benefits of economic growth have succeeded in keeping the number of single parents low. The narrow focus on definitions of the household and the family have also served to undermine single parents' position and free-doms in society. The consequences of such exclusionary practices, however, may hold greater consequences for certain women who live in abusive situations and are tied to violent relationships. Domestic violence is extensively reported in most nations regardless of religion, language, economic status or culture (Tomasevski 1993) and has now been recognised as an unacceptable violation of women's human rights (Heise 1995). In Singapore, the social stigma of single parenthood, exclusion and restrictions on housing and workforce participation serve only to confine women who are victims of domestic violence to a life of inequality, fear and abuse in the name of preserving the state's perception of a national ideology, values and morals. This represents one crucial area of gender relations in which the government avoids intervention and reverts to the notion of the household as a private domain.

CONCLUSION

Safeguarding and delivering equality, choice and other components of development and human rights for women is not possible if planning priority is given to
ensuring success, economic growth and racial harmony (Howard 1995). By pre-
scribing a specific selection of Asian cultural values and morals in Singapore and
implementing them through intrusive planning agendas the government has constructed national, ethnic and gendered identities and framed them by a series of
social norms and legal restraints. Others in Asia, however, do not share these
Asian cultural visions. The Malaysian Deputy Prime Minister, Anwar Ibrahim,
declared of Singapore, 'It is altogether shameful, if ingenious, to cite Asian values
as an excuse for autocratic practices and denial of basic rights' (*Independent on
Sunday* 19 March 1995). These values are part of a purely economic agenda, and
allow the preservation of an economic and material culture to take precedence
over equality and rights. They must be viewed as human rights violations.
Demands for cultural change and the recognition and respect of diversity and
difference must be met before a full realisation of women's rights can be achieved.
The most cited goal of planning and rights discourse is the empowerment of
women through control, capacity and respect (Karl 1995), but inherent in this,
Moser (1993) notes, is confrontation, not only with private actors but as in this
case with the state. Instead of being empowered women may find their equalities
and rights reduced. The action to take is inevitably difficult, but if women's rights
are to be respected as human rights, and the impact of planning on the security of
rights for women understood, the starting point must always be that recognised,
injustices against women are investigated and recorded as such. It is only then that
inequality, differential treatment and exclusion can begin to be addressed and
eliminated, and women can begin to emerge from the margins of development
and planning.

NOTES

1 This research is part of a larger doctoral research project on gender, urban poverty and
  vulnerability in Singapore. Research was conducted on low income households between
  1994 and 1996.
2 Almost all Singaporeans are of migrant descent. The current population of 2.7 million
  consists primarily of Chinese (78%), Malay (14%) and Indians (7%) (Department of
  Statistics, 1994).
3 The term 'family' is used throughout this chapter where 'household' would more commonly be used. This is to reflect current government rhetoric. Families may be nuclear
  or extended but are almost exclusively based on the notion of marriage.
4 O-Levels are secondary school examinations usually taken at the age of 15–16 years.
5 At the time of study $1 = S$1.5 or £1 = S$2.5.
6 One-parent households in Singapore are overwhelmingly headed by a divorced or
  widowed woman.
7 Average monthly household income in 1990 was calculated at S$3,076 (Department of
  Statistics, 1993).
8 Couples are defined as married, not co-habitating.

REFERENCES

Brown, D. (1994) *The State and Ethnic Politics in Southeast Asia*, London: Routledge.

Bunting, A. (1993) 'Theorising women's cultural diversity in feminist international human rights strategies', *Journal of Law and Society*, 20(1): 8–22.

Chan, H. C. (1989) 'The PAP and the structuring of the political system', in K. S. Sandhu and P. Wheatley (eds) *The Management of Success: The Moulding of Modern Singapore*, Singapore: Institute of Southeast Asian Studies.

Cheung, P. (1991) 'Commentary', in M. T. Yap (ed.) *Social Services: The Next Lap*, Singapore: Times Academic Press.

Chiew, Seen Kong (1990) 'National identity, ethnicity and national issues', in J. S. T. Quah (ed.) *In Search of Singapore's National Values*, Singapore: Times Academic Press.

Ching, F. (1994) 'Fay case: collision of values', *Far Eastern Economic Review*, 5 May.

Chun, A. (1996) 'Discourses of identity in the changing spaces of public culture in Taiwan, Hong Kong and Singapore', *Theory, Culture & Society*, 13(1): 51–75.

Davidson, G. M. (1996) 'The spaces of coping: women and "poverty" in Singapore', *Singapore Journal of Tropical Geography*, 17(2): 313–31.

Davidson, G. M. and Drakakis-Smith, D. (1997) 'The price of success: disadvantaged groups in Singapore', in C. Dixon and D. Drakakis-Smith (eds) *Uneven Development in Southeast Asia*, Aldershot: Avebury.

Department of Statistics (1994) *Yearbook of Statistics, Singapore 1993*, Singapore: Department of Statistics.

Devan, J. and Heng, G. (1994) 'A minimum working hypothesis of democracy for Singapore', in D. da Cunha (ed.) *Debating Singapore: Reflective Essays*, Singapore: Institute of Southeast Asian Studies.

Doran, C. (1996) 'Global integration and local identities: engendering the Singaporean Chinese', *Asia Pacific Viewpoint*, 37(2): 153–64.

Drakakis-Smith, D., Graham, E., Teo, P. and Ooi, G. L. (1993) 'Singapore: reversing the demographic transition to meet labour needs', *Scottish Geographical Magazine*, 109(3): 152–63.

Evans, R. (1990) 'Quest for an international legal code', *Geographical Magazine*, February: 34–7.

Financial Times (1995) *The Financial Times Survey 1994: Singapore*, 24 February 1995.

Goh, E. (1991) 'Service directions for voluntary welfare organisations serving disadvantaged families and children', in M. T. Yap (ed.) *Social Services: The Next Lap*, Singapore: Times Academic Press.

Gook, A. S. (1981) 'Singapore: a Third World fascist state', *Journal of Contemporary Asia*, 11(2): 244–54.

Heise, L. L. (1995) 'Freedom close to home: the impact of violence against women on reproductive rights', in J. Peters and A. Wolper (eds) *Women's Rights Human Rights: International Feminist Perspectives*, London and New York: Routledge.

Heng, R. (1994) 'Give me liberty or give me wealth', in D. da Cunha (ed.) *Debating Singapore: Reflective Essays*, Singapore: Institute of Southeast Asian Studies.

Hettne, B. (1996) 'Ethnicity and development: an elusive relationship', *Contemporary South Asia*, 2(2): 260–74.

Howard, R. E. (1995) 'Women's rights and the right to development', in J. Peters and A. Wolper (eds) *Women's Rights Human Rights: International Feminist Perspectives*, London and New York: Routledge.

Jones, G. W. (1984) 'Introduction', in G. W. Jones (ed.) *Women in the Urban and Industrial Workforce: Southeast and East Asia*, Developing Studies Centre, ANU, Monograph No. 33, Canberra.

Kabeer, N. (1995) *Reversed Realities: Gender Hierarchies in Development Thought*, London: Verso.

Karl, M. (1995) *Women and Empowerment: Participation and Decision Making*, London: Zed Books.

Kaufman, N. H. and Lindquist, S. A. (1995) 'Critiquing gender-neutral treaty language: the convention on the elimination of all forms of discrimination against women', in J. Peters and A. Wolper (eds) *Women's Rights Human Rights: International Feminist Perspectives*, London and New York: Routledge.

Kodras, J. E. and Jones, J. P. (1991) 'A contextual examination of the feminization of poverty', *Geoforum*, 22(2): 159–71.

Kong, L. (1995) 'Music and cultural politics: ideology and resistance in Singapore', *Transactions of the Institute of British Geographers* 20: 447–59.

Massey, D. (1994) *Space, Place and Gender*, Cambridge: Polity Press.

Ministry of Community Development (1995) *Singapore: A Pro-Family Society*, Singapore.

Ministry of Community Development (1996) *Helping Low-Income Families: The Singapore Way*, Singapore.

Moser, C. O. N. (1993) *Gender Planning and Development: Theory, Practice and Training*, London and New York: Routledge.

National Advisory Council on the Family and the Aged (1995) *Family Values: Singapore*, Singapore: Ministry of Community Development.

Peters, J. and Wolper, A. (eds) (1995) *Women's Rights, Human Rights: International Feminist Perspectives*, London and New York: Routledge.

Pyle, J. L. (1994) 'Economic restructuring in Singapore and the changing roles of women, 1957 to present', in N. Aslanbeigui, S. Pressman and G. Summerfield (eds) *Women in the Age of Economic Transformation: Gender Impact of Reforms in Post-Socialist and Developing Countries*, London and New York: Routledge.

Quah, J. S. T. (ed.) (1990) *In Search of Singapore's National Values*, Singapore: Times Academic Press.

Quek, L. T. A. (1992) 'Single parents and informal networks in Singapore', unpublished masters thesis, Department of Sociology and Social Work, National University of Singapore.

Rao, A. (1995) 'The politics of gender and culture in international human rights discourse', in J. Peters and A. Wolper (eds) *Women's Rights Human Rights: International Feminist Perspectives*, London and New York: Routledge.

Rigg, J. (1991) *Southeast Asia: A Region in Transition*, London: Unwin Hyman.

Rodan, G. (1993) 'Preserving the one-party state in contemporary Singapore', in K. Hewison, R. Robison and G. Rodan (eds) *Southeast Asia in the 1990s: Authoritarianism, Democracy and Capitalism*, London: Allen and Unwin

Ruddick, S. (1996) 'Constructing difference in public spaces: race, class, and gender as interlocking systems', *Urban Geography* 17(2): 132–51.

Saw, S. H. (1990) 'Changes in the fertility policy of Singapore', Occasional Paper No. 3, Institute for Policy Studies, Singapore.

Sibley, D. (1992) 'Outsiders in society and space', in K. Anderson and F. Gale (eds) *Inventing Places: Studies in Cultural Geography*, Melbourne: Longman Cheshire and Wiley.

Singham, C. (ed.) (1996) 'Introduction', in Association of Women for Action and Research (AWARE) (ed.) *The Ties That Bind: In Search of the Modern Singapore Family*, Singapore: AWARE.

Soin, K. (1996) 'National policies – their impact on women and the family', in AWARE (ed.) *The Ties That Bind: In Search of the Modern Singapore Family*, Singapore: AWARE.

Straits Times (Singapore), various issues.

Sullivan, D. (1995) 'The public/private distinction in international human rights law', in J. Peters and A. Wolper (eds) *Women's Rights Human Rights: International Feminist Perspectives*, London and New York: Routledge.

Teo, P. and Ooi, G. L. (1996) 'Ethnic differences and public policy in Singapore', in

D. Dwyer and D. Drakakis-Smith (eds) *Ethnicity and Development: Geographical Perspectives*, Chichester: John Wiley.

Tomasevski, K. (ed.) (1993) *Women and Human Rights*, London: Zed Books.

United Nations Development Program (UNDP) (1995) *Human Development Report 1995*, New York and Oxford: Oxford University Press.

Vaiou, D. (1992) 'Gender divisions in urban space: beyond the rigidity of dualist classifications', *Antipode* 24(4): 247–62.

Varley, A. (1996) 'Women heading households: some more equal than others?', *World Development* 24(3): 505–20.

Willmott, W. E. (1989) 'Emergence of nationalism', in K. S. Sandhu and P. Wheatley (eds) *The Management of Success: The Moulding of Modern Singapore*, Singapore: Institute of Southeast Asian Studies.

Wolf, D. (1990) 'Daughters, decisions and domination: am empirical and conceptual critique of household strategies', *Development and Change* 21: 43–74.

Wong, A. (1981) 'Planned development, social stratification and the sexual division of labour in Singapore', *Sign: Journal of Women in Culture and Society*, 7(2): 424–52.

World Bank (1991) *World Development Report 1991*, New York: Oxford University Press.

World Bank (1993) *The East Asian Miracle: Economic Growth and Public Policy*, New York: Oxford University Press.

Part III

GENDER, DEVELOPMENT AND POLICY-MAKING WITHIN THE HUMAN RIGHTS CONTEXT

6

HOUSEHOLDS, VIOLENCE AND WOMEN'S ECONOMIC RIGHTS

A case study of women and work in Appalachia

Ann M. Oberhauser

INTRODUCTION

The United Nations Decade for Women (1976 to 1985) increased awareness of the differential impact of development on women and men and strengthened the agenda for advancing and reconceptualizing women's human rights. In 1995 the United Nations Fourth World Conference on Women called for continued efforts to improve women's socio-economic status and reflect on progress in the area of women's human rights. During these two decades, development strategies to enhance women's positions have changed significantly. Initially, the women-in-development approach stated that ending sex discrimination would increase over-all economic efficiency, and thus contribute to economic development. More recently, development efforts have shifted from examining women in isolation to analyzing how gender relations and divisions of labor are related to social, spatial and economic inequalities in Third-World regions (Folbre 1995; Moser 1993; Tinker 1990).

The reconceptualization of women, gender and development since the late 1970s has also had an impact on human rights formulations. Feminists argue that it is not enough for human rights to be merely extended to women; abuses that stem from gender-based discrimination must be considered human rights violations (Binion 1995). This redefinition of human rights standards entails shifting the focus from the public to the domestic sphere, where violations of women's rights are most likely to take place. Moreover, by re-examining the notions of "public" and "private" in human rights discourse, one can better understand the significance of unequal opportunities for women in education, health and employment (Peters and Wolper 1995).

This chapter examines the relationship between gender dynamics in the household and women's unequal access to employment opportunities in the semi-peripheral region of Appalachia. Underdevelopment in both the First and Third Worlds usually includes a history of raw material extraction and broad socio-economic structures of exploitation that contribute to local resistance and struggle. Appalachia is often depicted as a peripheral region, in a core industrialized country, where women represent an especially marginalized population (Oberhauser 1995a; Pudup *et al.* 1995). This chapter argues that analyses of intra-household gender dynamics offer considerable insight into the discrimination women face in accessing resources and improving their economic status, particularly in economically depressed regions. Two themes from the gender and development literature support this argument. First, dominant gender relations and divisions of labor in the household contribute to social and economic inequalities between women and men. Conventional approaches to development and traditional western views of human rights tend to make broad, generalized assumptions about the household as a homogenous, unified unit where members have equal control over resources and decision-making powers. In contrast, feminist analyses view the household as a heterogeneous unit with complex and often conflictive power dynamics (Benería and Roldán 1987; Folbre 1986; Moser 1993).

A second theme in this discussion is women's empowerment through engagement in collective efforts at the community level. Community-based organizations that mobilize women and other marginalized groups around economic and social rights have strengthened the struggle for women's empowerment and human rights. Women's rights organizations, for example, argue that unequal opportunities in education, health and employment are linked to gender-based discrimination and abuse. This link is especially relevant to women in economically depressed regions of both the First and Third Worlds.

Theoretical and empirical issues relating to women's rights, intra-household relations, and gender and development are explored here in three sections. The first section challenges conventional development literature for its conceptualization of the household as a harmonious unit separate from the public sphere. The discussion offers an alternative approach that examines gender inequalities in the household and analyzes the relationship between the private and public spheres. It is argued that women's access to economic opportunities and decision-making is affected by the simultaneous and often overlapping productive and reproductive roles of women. This section also discusses how intra-household dynamics involving gender-based violence are often linked to women's marginal economic status (Moser 1993; Peters and Wolper 1995). Throughout this discussion, the lower status of women is examined in the context of their gender, race, class and geographical context.

The second section applies this theoretical analysis to Appalachia, where gender relations and household strategies have been largely neglected in economic development planning. Gender-based abuses in this region are linked to traditional attitudes towards women, especially their domestic roles, the prevalence of violence towards women, and the inability of many women to gain access to well-paid and secure employment. The final section outlines a development approach that employs community-based strategies to advance women's socio-economic status. The proposed strategies build on existing social and economic networks in order to secure a gender-, class- and racially-balanced regional development process.

INCORPORATING HOUSEHOLD GENDER RELATIONS IN DEVELOPMENT AND HUMAN RIGHTS

Conventional approaches to economic development in both First- and Third-World settings tend to use national and regional statistics on income, poverty, employment and health care to examine socio-economic trends and implement policy. These approaches generally ignore community and household dynamics, or depict the household as a broad, generalizable category whose members have equal control over resources and whose income is primarily generated by male "breadwinners." Consequently, intra-household inequalities and human rights abuses in the domestic sphere are often neglected. Moreover, efforts to reduce socio-economic disparities overlook the potential for community-based economic development.

In contrast, gender and development approaches have for decades recognized the heterogeneity of households (Rogers 1980), the significant proportion of women who are *de facto* heads of household (Brydon and Chant 1989), and the unequal access to resources within households (Dwyer and Bruce 1988; Folbre 1995). These analyses are important in that they demonstrate how families and households are often conflictive, differentiated social and economic units where decisions are frequently made on the basis of unequal power relations. Examining gender relations in the household is also an important way of linking gender-based discrimination against women in the domestic sphere and their marginal status in the workplace and public domain. According to Peters and Wolper (1995), the subordination of women in the household should be considered a human rights abuse because it contributes to their marginalization in low-wage occupations and, for some, an inability to engage in income-generating activities altogether.

The relationship between gender inequalities in the household and workplace relates to the false dichotomy often constructed between public and private, production and reproduction, workplace and domestic spheres, and, ultimately, masculine and feminine (Dwyer and Bruce 1988; Moser 1993). Many feminists claim that this dichotomy is invoked to justify women's subordination in both the household and workplace and excludes human rights abuses in the home from public scrutiny (Peters and Wolper 1995). In their work on state responsibility under international law, Thomas and Beasley (1993: 40) argue that

> Nowhere is the effect on international human rights practice of the public/private split more evident than in the case of domestic violence – which literally happens "in private." States dismiss blatant and frequent crimes, including murder, rape, and physical abuse of women in the home, as private, family matters, upon which they routinely take no action.

Gendered approaches to human rights are also useful in comparing development processes and planning in diverse geographical contexts. For example, many women in peripheral regions of the industrialized North have experienced similar forms of marginalization and unequal access to resources as women in developing regions of the South. Additionally, development of these peripheral areas has generally been approached in ways that ignore household gender relations and community-based economic strategies.

Third-World feminists challenge both conventional development theory and western feminism for producing particular ethnocentric discourses that tend to universalize Third-World women and neglect their historical and material hetero-geneity. The varied experiences and backgrounds of marginalized women in both the First and Third Worlds underline the necessity to allow "other" voices to speak in challenging the oppressive power structures of western, white, hetero-sexual and masculinist discourses. Likewise, the example of poor white women in economically depressed, rural regions such as Appalachia illustrates the import-ance of examining women's marginalization in their particular historical and material contexts. According to Mohanty (1991), Third-World women include marginalized women in both developing and developed countries who are engaged in a common struggle against exploitative structures. Many feminists critique the representation of Third-World women as poor, uneducated, and tradition-bound, and instead encourage approaches and strategies that reflect the historical, cultural and material complexities of their lives in addition to their position as active agents of change (Benería and Roldán 1987; Dwyer and Bruce 1988).

In contrast to conventional development approaches that operate within west-ern norms, women's human rights approaches incorporate the cultural and his-torical context of societies in developing countries. For example, debates over genital mutilation, feet-binding, purdah and sex trafficking often fail to recognize the cultural context and the structural interrelatedness of the various manifest-ations of gender discrimination (Mohanty *et al.* 1991; Peters and Wolper 1995). Third-World feminism is useful in this analysis of gender, development and human rights because it demonstrates that the marginalization of women is closely related to the social construction of gender, race and class in particular historical and material contexts.

Household dynamics and women's socio-economic status

Both Western and Third-World feminists address the relationship between women's subordinate positions in the household and their reduced economic opportunities (Benería and Stimpson 1987; Dwyer and Bruce 1988; Seitz 1995). The discussion below examines this relationship from a human rights perspective: first, by analyzing how women's primary role as caretakers limits their access to gainful employment and second, by examining how domestic violence psycho-logically and physically affects women's ability to improve their socio-economic status. Global data on the socio-economic status of women and violence toward

women show that unequal household relations and women's economic dependence on men is a manifestation of gender-based discrimination and thus a
violation of women's human rights.

In agricultural societies, especially those in developing regions, distinct gender
divisions of labor help determine access to and control over resources (Folbre
1995; Tinker 1990). Women generally perform work in the fields and as well tend
to the daily needs of their families. In more developed and industrialized regions, a
significant proportion of women engage in paid labor, but still have more household responsibilities than men. These divisions of labor by gender often limit
educational opportunities and employment training for women. For example,
many women and girls in developing countries are discouraged from obtaining
university degrees or even completing their secondary education because these
activities are seen as unnecessary in preparing for their domestic roles (Brydon and
Chant 1989). One measure of the relative lack of education among girls and
women is illiteracy rates, which are consistently higher among women than men
throughout the world. Women in rural areas are particularly disadvantaged.
According to the United Nations (1995), illiteracy rates among rural women in
some countries are two to three times higher than in urban areas. This situation is
partly due to the prevalent attitude that girls should assume roles that do not
require formal education, such as preparing for marriage and having children.

Women's primary role in the domestic sphere and their unequal access to
education affect their position in the paid labor force. Although women make up
nearly half the workforce in most developed economies, they are not evenly
distributed among occupations and sectors. The majority of women in these
regions are employed in non-professional, lower status occupations such as clerical, sales, service or factory jobs, which are lower paid and less secure positions.
Less than one-fifth of all female workers are in higher status, male-dominated
occupations (those in which 75 per cent or more of the workers are men) (UN
1995). In sum, household gender relations and divisions of labor influence the
inequalities women face in education and employment opportunities and must be
considered in human rights discussions.

The second issue linking women's subordinate roles in the domestic sphere and
their unequal access to employment is the physical and emotional abuse suffered
by many women in the household. The recurrent abuse and violence towards
many women in the privacy of their homes challenges the myth of the domestic
sphere as a harmonious unit and safe haven (Thomas and Beasley 1993).
Although gender-based violence is an abuse of human rights, violence in the
private sphere of the household is often overlooked by governments and even the
human rights community because it does not occur within the realm of state-
sanctioned oppression in the public sphere (Bunch 1990; Moser 1993). The
majority of domestic violence, the most prevalent violation of women's rights
throughout the world, is reported to be abuse by a husband or intimate partner
(UN 1995). Studies in ten countries estimated that between 17 and 38 per cent of
women have been physically assaulted by an intimate partner. Cross-cultural

97

comparisons indicate that Africa, Latin America, and Asia tend to have the highest rates, with up to 60 per cent of women suffering physical abuse (UN 1995). Other forms of physical violence include the sexual control of women's bodies such as rape, forced prostitution, sexual assault, genital mutilation, or the beating of girls and women.

Violence towards women and emotional abuse is psychologically degrading and can be traced to institutional forms of racial and gender discrimination. Domestic violence often leads to loss of self-confidence, which may prevent women from participating fully in the workforce. In both industrialized and developing countries, for example, domestic violence and rape cause women to lose a significant percentage of work days (World Bank 1995). Among Third-World women, gender-based violence often has important historical roots. Mohanty (1991) identifies a definite project of sexual exploitation that was imposed by colonial rule in developing countries whereby indigenous women were abused or even enslaved. Similarly, African women brought to the Americas as slaves were frequently beaten, raped, and impregnated by their white masters (Morrisey 1989). In many of these examples, abusive behavior can be explained as what men do to prevent women from obtaining economic power. This show of physical force or emotional abuse is an important, yet often neglected, human rights abuse.

Gender inequality and women's economic strategies

Despite these limitations on women's participation in the paid labor force, women throughout the world play an important part in household and community economic strategies. The income they generate from both formal and informal sector activities is vital to the well-being of families and communities (Moser 1993; Smith and Wallerstein 1984; World Bank 1995). Indeed, studies indicate that women generally devote a far larger share of their income and earnings to family needs such as child health and nutrition than men do (Benería and Roldán 1987; Folbre 1995). Community-based economic activities are a common form of income-generation among women, especially in peripheral regions where formal employment opportunities are less available. Although these activities can contribute to important networking and collaboration among women, the products and services are often undervalued and suffer from lack of adequate markets (Oberhauser 1995b).

Economic restructuring in many industrialized regions has contributed to women's increasing role in household income generation. Throughout these regions, the loss of jobs in male-dominated occupations has paralleled a rise in female-dominated jobs in the service sector (Benería and Stimpson 1987). Two aspects of this trend, however, reinforce gender inequality in the economic sphere. First, women's earnings in service sector jobs have not made up for the loss of earnings from male-dominated jobs, partly because a large proportion of service sector jobs are low paid and part time. Second, although the percentage of women 15 years old and over who are economically active is increasing, these rates lag behind men's participation in the paid labor force. In the United States, female economic activity rates increased from 43 per cent in 1970 to 58 per cent in 1990 while male activity rates actually decreased from 79 per cent to 76 per cent in that same period (US Bureau of Labor Statistics 1990). On a global scale, Latin America has some of the lowest economic activity rates for women with 34 per cent in 1990 compared to 82 per cent of men (UN 1995). These figures are misleading, however, because they do not account for income-generating activities outside the realm of formal waged employment. In regions such as Latin America, informal activities such as crafts production and street-vending among women contribute greatly to household economies (Benería and Roldán 1987).

Although their participation in both formal and informal economic activities outside the home is increasing, women continue to assume major responsibility for household work and the care of children and other family members (Folbre 1995; Tinker 1990). Time-budget surveys indicate that women work more hours than men when economic and non-economic activities are included. Thus women are often faced with the difficult task of juggling the triple roles of reproductive, productive and community work (Moser 1993; UN 1995).

This brief overview of women's participation in both paid and unpaid labor indicates that the subordination of females in the household and workplace contributes to their lack of basic economic rights. Therefore, gender analyses of household dynamics can greatly inform our understanding of development and human rights issues in both Third- and First-World contexts. This discussion, challenging the assumption that households are homogenous, unified structures, encourages a critical analysis of the often conflictive and dynamic nature of gender relations and divisions of labor in the household. It is argued that unequal access to resources and unequal distribution of power in this area contribute to human rights abuses such as gender-based domestic violence and limited access to economic opportunities. The following section applies these gender and development themes and human rights issues to household dynamics and gender inequalities in Appalachia.

GENDER AND DEVELOPMENT IN APPALACHIA: FROM THE THIRD WORLD TO THE FIRST AND BACK AGAIN

Feminist analyses of development and human rights is relevant to Appalachia where social, economic and gender inequalities can be as pronounced as those in the Third World. This section examines gender and development in the Appalachian region in the Unites States, with a particular focus on the state of West Virginia. The discussion begins with a brief overview of different theoretical approaches to Appalachian regional development, to underline the interdependence of underdevelopment in both First- and Third-World contexts. The second part outlines intra-household dynamics and gender relations as a means of understanding social inequality and economic marginalization of women in this region. The consequences of women's primary responsibility for domestic tasks and the relatively high rate of domestic violence in West Virginia are examined as human rights issues that contribute to women's unequal access to economic opportunities.

Theorizing Appalachian development

Appalachia is often depicted as a peripheral region that has been negatively affected by external economic and political forces, thus deprived of its self-determination (Fisher 1993; Pudup *et al.* 1995; Walls 1976). Approaches to under-development in this region can be categorized into three groups: the culture of poverty approach, the internal colony model, and collective resistance. Early explanations of underdevelopment in Appalachia include cultural stereotyping of the local people as lazy and backward (Weller 1965). According to this approach, the perpetuation of Appalachia's underdevelopment is a result of subcultural norms that consign the population to an underprivileged status. In order to move out of this backward status, the region must modernize in accordance with set stages of capitalist development.

A second explanation of Appalachian underdevelopment shifts the discourse to broader socio-economic structures which place the region in a Third-World pos-ition, one of being exploited by external capital. The internal colony model places Appalachia as an exporter of raw materials to the developed core (Walls 1976). More recently, Appalachian scholars and activists have challenged the misleading stereotypes of Appalachians as victims or passive recipients of cultural norms or external exploitation. This contemporary approach emphasizes the economic and social diversity in this region and the importance of connecting regional devel-opment to national and global economic restructuring (Fisher 1993). Instead of portraying Appalachian people as inferior and victims of exploitation, they focus on the resistance and struggle that have shaped this region's economic and social history (Gaventa *et al*. 1990).

Appalachian studies has traditionally focused on male-dominated formal sector activities and ignored women's economic activities and crucial reproductive labor. Feminists have challenged the androcentrism that pervades many of these studies and provided alternative analyses of gender in Appalachian regional and eco-nomic development (Anglin 1993; Maggard 1994; Oberhauser 1995b; Pudup 1990; Smith 1995). This study builds on gender analyses of historical and con-temporary economic restructuring in Appalachia better to understand the con-nection between household gender relations and women's human rights. The analysis draws on an intensive study by the author on women's home-based work in Appalachia and research by Oberhauser, Waugh and Weiss (1996) on gender and economic development in West Virginia.

A gender analysis of economic development in West Virginia

The development of Central Appalachia, and West Virginia in particular, is linked to the area's historical economic geography as well as dominant gender relations and divisions of labor in the household and workplace. Figure 6.1 illustrates the rural nature of West Virginia and its position within the broader Appalachia region. Central Appalachia's rich natural resources began to attract outside capital on a large scale at end of the nineteenth century (Lewis 1993; Pudup 1990). During this early period of industrialization, firms from outside the region invested in the timber industry, coal mining and natural gas extraction, often realizing large profits in previously unexploited territory. Communities throughout West Virginia experienced rapid increases in their populations and economic activity as migrants from Europe and the South came to work in the expanding industries (Lewis 1993).

Gender relations and divisions of labor were critical components of this early period of industrial development. Productive and reproductive labor was clearly divided along gender lines (with some important exceptions) as men worked in the mines, mills, and factories and women were primarily involved in household activities such as cooking, cleaning, childcare, and other domestic work (Greene 1990). In addition to these reproductive activities, some women contributed to their household incomes by taking in boarders, selling garden produce, or providing services such as washing laundry or childcare. Other women were employed outside the home in textile and knitting mills, glassware shops, and tobacco factories (Hensley 1990).

The rural geography of West Virginia, alongside the gender-specific employment opportunities, partially explain the marginalization of women in the labor force. Male employment in lumbering and coal mining was abundant in transient boom towns or in labor camps, while female employment opportunities were in short supply. Access to formal, waged employment, however, varied among women in the state according to their racial and ethnic background (Pudup 1990). For example, many black and immigrant women who moved to urban areas in the state in search of wage labor found work in factories, inns, private houses and laundry services. In contrast, white, native-born females who comprised the majority of West Virginia women tended to remain in their rural communities and did not seek employment outside the home. Overall, women's primary role in the reproductive realm of the household has never precluded them from engaging in economic activities to help support their households. Income-generating activities primarily entailed informal work such as taking in boarders, cleaning houses, and selling produce, but also included some formal sector employment. During this early period of industrialization, the percentage of women in the workforce rose from 6 per cent in 1870 to 9.3 per cent in 1900 (WV Bureau of Employment Programs 1993).

The almost total exclusion of women from paid labor during the early period of industrialization in West Virginia has contributed to the contemporary

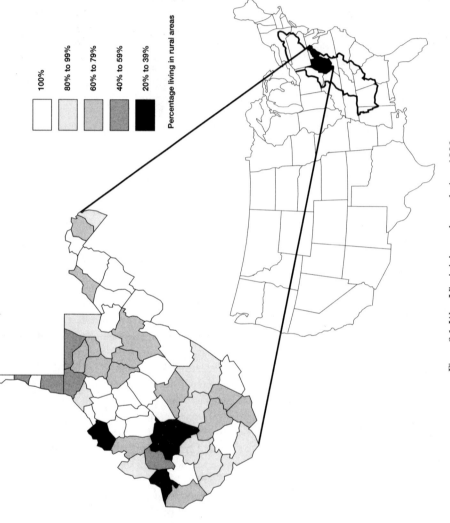

Figure 6.1 West Virginia's rural population, 1990
Source: US Census 1990

marginalization of women in the workforce. Recent statistics indicate that West Virginia women lag behind WV men and their national counterparts in several areas. Disparities between men's and women's wages and labor force participation exist at both the state and national levels (see Figure 6.2). Nationally, women earn 60 cents for every dollar earned by men. The wage gap for WV women is even greater as they earn 45 cents to every dollar earned by men in West Virginia. In addition, in 1993, the national women's labor force participation rate, or the percentage of women 16 years and older who are economically active, was 57 per cent compared to 76 per cent for men. In comparison, only 43 per cent of working-age women and 65 per cent of working-age men were employed in WV (US Bureau of the Census 1993; WV Bureau of Employment Programs 1993). The relatively low wages and the low percentage of women employed is related to their greater poverty (Oberhauser *et al.* 1996). As illustrated in Figure 6.2, 21 per cent of West Virginia women live in poverty (below the general level of income set by government as adequate) compared to 15 per cent of women nationally (US Bureau of the Census 1993). These data reveal the economically disadvantaged position of West Virginia women, which is partially explained by the pronounced gender roles within household, as outlined below.

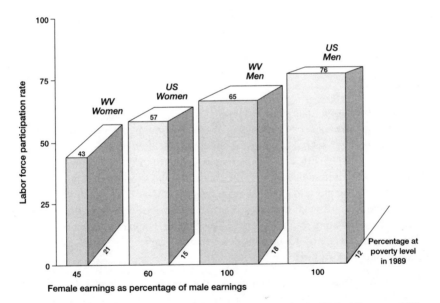

Figure 6.2 A gender comparison of employment and poverty in the United States and West Virginia, 1990 and 1993

Sources: West Virginia Bureau of Employment Programs, 1993; US Bureau of Census 1993

Household gender relations and domestic violence in Appalachia

As argued above, the unequal positions of Third-World men and women within the domestic sphere contribute to their unequal control over resources (Moser 1993). These same issues apply to households in peripheral regions of the industrialized North such as Appalachia. Development in this region is examined here through an analysis of the power relations and resource allocation within households, as well as uneven development at the regional scale. Specifically, traditional attitudes about women's roles in the household and domestic violence deny Appalachian women equal employment opportunities, and thus violate their basic human rights. The discussion draws from my own research, in which I conducted semi-structured interviews with over eighty homeworkers in West Virginia.

Similar to the gender roles of Third-World women outlined above, the roles of girls and women in Appalachia are guided by the culture and socio-economic forces in general. They have put women on the margins of paid labor and made them primarily responsible for reproductive work. Under prevalent gender roles and divisions of labor in Appalachia women assume the major responsibility for domestic tasks, marrying early and providing supplementary income for the household when necessary (Seitz 1995). One of the women who participated in the author's study confirmed this general attitude towards women in West Virginia:

> I have lived in New Jersey, Virginia, Maryland, Pennsylvania, and I have found that WV is so backwards in its thinking. "Women should be home, women should be with the babies, tending the house, women shouldn't be working" – except if they put 'em in as a waitress at Geno's [local restaurant].

This perception of female roles has limited women's access to careers and education and the likelihood of advancing them. Although the percentage of WV females with high school diplomas is higher than that of WV males, the percentage of females with college degrees is lower than the percentage of males (Hannah 1995). These figures indicate that girls who finish high school are often discouraged from attending college or getting any advanced education.

In addition, traditional attitudes towards women not only limit their efforts to gain further education, but also to establish a career. For example, Mary's decision to postpone her plans to gain a degree after starting a family affected her employment opportunities:[1]

> I worked in restaurants, probably a year, and got married when I was 19. I had two children and decided to go back to school because I wanted to teach, but got pregnant in the middle of the year. Since I never left the children to go out to work, I says well, that does that, no more school for me. When my daughter started school, I decided to go to business college, so I went to business college. Then I did bookkeeping sixteen years before I retired.

Mary's situation illustrates how advanced education and careers for women are often interrupted by family responsibilities that include marriage and having children. In general, WV women are more likely to be married than American women in general and tend to marry younger. Among 15–19 year old females in WV, 7 per cent were married compared to only 5 per cent of US women (US Bureau of the Census 1990). In addition, major differences in marital patterns exist among racial groups in WV (Hannah 1995). In 1990, black women were almost twice as likely to be single as white women. Additionally, fertility rates among young women aged 20–24 in WV were higher than those for the US: 121 in WV compared to 117 births per 1,000 among women in the US (National Center for Health Statistics 1993). Given the relatively high teenage pregnancy rate, many schools are beginning to provide childcare and offer child development courses (Hannah 1995).

One of the major domestic responsibilities among women in West Virginia is caregiving, defined here as the provision of services to those unable to care for themselves, such as the elderly, children, and the chronically disabled or ill. Families with adult wage earners working outside the home often have the greatest need for caregiving services. Women are most likely to provide care to those in need if caregiving services are not available or affordable. Violet's story illustrates how caregiving responsibilities for needy family members entail significant sacrifices and limit a womans's ability to engage in formal employment:

> My aunt, who is 64 years old, is handicapped. She's deaf and uses sign language and she's lived in our household for twelve years. Recently she's had some congestive heart failure complications and it's necessary for me to be with her a good bit, to at least run in and check on her. She can take care of her independent needs right now, but her outlook is not very good as far as long-term care, so I never know when she might go to the hospital. Because of her communication disorders, I have to be with her at the hospital most of the time and that would be very disruptive to make a commitment working outside the home.

Like many other West Virginia women, Violet lives in an area where health care and caregiving services are somewhat inaccessible or unaffordable for a significant proportion of the population. As the ageing of the population in West Virginia continues to outpace the national rate, women will increasingly take on caregiving roles as parents live to a greater age than formerly (Oberhauser et al. 1996). Indeed, demographic trends indicate there will be a greater need for respite care and social support networks as the number of women who provide significant time to care for the elderly increases. According to Oberhauser et al. (1996: 7–8):

> These caregiving responsibilities translate to tremendous sacrifices in the area of formal, waged employment. It is estimated that over a lifetime, women lose eleven years of wage-labor productivity to caregiving, especially caring for their children. These responsibilities frequently amount to a

full-time (and often second) job for which they typically receive no compensation.

Childcare is also a major constraint for women seeking to enter the workforce. Approximately 46 per cent of children under age 6 in WV required some kind of non-parental care in 1990 (Hannah 1995). Three-quarters of parents who seek day care do so because of employment needs. In many cases, other family members or neighbors provide the much-needed childcare, which sometimes raises concerns about the quality and reliability of these situations. Lower-income families are disproportionately affected by childcare costs, yet are also most in need of the services and financial support provided by state and federal programs. Childcare consumes up to one-quarter of the income of a family living at the poverty level, but only 6 per cent of upper-income family resources (Poersch *et al.* 1994). In 1994, West Virginia's demand for federally-subsidized childcare slots outstripped available funds, putting over 2,000 children on the waiting list for childcare and placing severe economic burdens on parents. Moreover, 29 per cent of the respondents to a recent survey of West Virginia's Aid to Families with Dependent Children and food stamp recipients indicated that lack of childcare was a barrier to their employment (Children's Policy Institute of WV 1995).

The lack of attention to women's primary role in domestic responsibilities such as caregiving is only one of the impediments to women's access to economic opportunities. In addition, Appalachian women trapped by domestic violence have limited access to the social support and material resources needed to seek gainful employment. As discussed above, domestic violence and sexual abuse is an extreme and insidious manifestation of gender inequality and women's subordination. Because it occurs in the private sphere of the home, it is frequently "invisible" and is thus considered outside the public domain.

West Virginia law has increased its ability to intervene in cases of violence towards women; however, the problem remains serious. In 1992, 91 per cent of identified domestic violence in West Virginia was directed against women. Moreover, from 1990 to 1992 reports of domestic violence increased by 98 per cent (from 3,040 to 6,029) (Hannah 1995). Evidence of the significance of the problem is the fact that severe abuse is the most likely reason for women to seek medical attention and in 1992, over one-third of murders in West Virginia were the result of domestic violence (Hannah 1995). Seitz addresses the physical oppression women face at the hands of male partners or husbands in her study of women and empowerment in south-west Virginia. The following incident involving physical violence is an example of the brutal and contemptuous attacks carried out by males in attempts to perpetuate their control and power (Seitz 1995: 59):

> We went out to dinner and he didn't have any change for the tip, so I reached in my purse and said: "Here honey" and laid the change down. As we got outside the door he said: "You won't belittle me like that!" I didn't understand what he was talking about! And then with that, he pushed me

against the car. I was seven months pregnant and the buttons popped off the front of my coat. It wasn't the first time.

This quote illustrates that, as elsewhere, violence towards women in Appalachia is linked to their subordinate position and is provoked by evidence of social changes that put control of women's lives and bodies in their own hands (Bunch 1990).

As more and more women increase their economic activities in West Virginia, male domination and control over household resources are threatened. As a result, male partners sometimes sabotage or physically retaliate against women who upgrade their skills and become employed outside the home. For example, as some women complete job training or enter the workforce, verbal harassment or even violence is inflicted by abusers, in order to demean their skills and aspirations. These incidents are all too common among women in West Virginia's thirteen domestic violence shelters, which provided a 20 per cent increase in services to victims of violence between 1991 and 1995 (WVCADV 1996).

In sum, the overriding attitudes towards gender roles and divisions of labor in the household and the gender-based violence and emotional abuse clearly impede women's access to economic opportunities; thus they represent important women's rights issues. This section has discussed these themes in the context of Appalachia and particularly the economic development of West Virginia. Throughout the discussion, the household is presented as a critical scale of analysis from which to evaluate and understand development processes. Subordination of women in the household and their exposure to domestic violence are barriers to the advancement of women that must be dealt with as human rights issues. Several strategies for improving women's socio-economic status are discussed in the concluding section.

CONCLUSION: FROM HOUSEHOLDS TO COMMUNITY-BASED ECONOMIC STRATEGIES

This chapter applies a gender analysis of development and human rights to an empirical study of the impact of household gender relations and divisions of labor on women's access to economic opportunities. The theoretical shift in the development and human rights literature from focusing on women's issues to analyzing gender as a basis for discrimination and abuse emphasizes how gender inequalities and gender-based human rights violations are increasingly recognized as important dimensions of societal structures and institutions. Gender and development studies have advanced our understanding of the link between women's economic status and human rights, arguing that unless women's status is improved, certain households, communities, and regions will continue to lag behind. This chapter contributes to this discussion by analyzing how gender inequalities in the household are linked to women's unequal access to economic opportunities. Specifically, predominant gender roles and divisions of labor in the household affect women's economic rights partly because they limit women's access to education and employment. Household dynamics are also related to gender-based domestic violence, which is often neglected in state-sanctioned forms of human rights oppression in the public sphere. Thus the household is an important site that needs to be incorporated in analyses of gender, development and human rights.

The focus on intra-household gender dynamics is a logical point from which to begin to implement community-based economic strategies. Development planners should build on analyses of household gender relations and divisions of labor to explore alternative means of income-generation, especially among women in low-income households. Several gender-sensitive strategies that stem from local initiatives have proven effective, especially in many Third-World contexts (Moser 1993). Indeed, numerous studies have argued that substantial gains can be made if economic development includes investment in and support for small-scale commercial efforts that target the economically and socially disadvantaged in poor regions (Dwyer and Bruce 1988; Gaventa et al. 1990; Tinker 1990). Specific examples of community-based projects such as micro-enterprise development, flexible manufacturing networks, and rotating loan programs are often successful in Third-World countries and are also applicable in the First World.

In many rural communities, the informal sector is a common economic strategy that generates household income through selling or bartering goods and services. Gender is an important factor in analyzing who and what is involved in this economic strategy. Women tend to be active in the informal sector in ways that relate to their domestic roles and household divisions of labor. They are also marginalized from formal activities, owing to factors outlined above. Therefore, common income-generating strategies among Appalachian women are selling hand-crafted items and agricultural produce or providing services such as care for the elderly and childcare or housecleaning (Oberhauser 1995a). Most of these activities exist outside the formal labor market, and thus are ignored by economic development efforts and marketing programs. Widely recognized and respected community-based organizations and grass-roots networks can avoid the exploitative nature of informal sector activities (Weiss 1990). Through community-based and gender-sensitive planning, women can learn valuable financial, marketing, and production skills that will greatly improve their economic status. These gains can sometimes translate to psychological empowerment that will provide the impetus for some women to get out of abusive situations and become more economically independent.

In sum, the incorporation of gender analyses in development and human rights debates emphasizes the important role of household gender relations in the distribution of resources among men and women and their access to economic opportunities. Gender planning represents a move toward collective strategies controlled by community groups that can empower women and provide long-term development for peripheral regions. These strategies are sensitive to household relations and divisions of labor, and will ultimately improve the socio-economic status of marginalized populations.

NOTE

1 The names of the interviewees have been changed in the text to ensure their confidentiality.

REFERENCES

Anglin, M. (1993) 'Engendering the struggle: women's labor and traditions of resistance in rural southern Appalachia', in S. L. Fisher (ed.) *Fighting Back in Appalachia*, Philadelphia, PA: Temple University Press.

Benería, L. and Roldán, M. (1987) *The Crossroads of Class and Gender*, Chicago: University of Chicago Press.

Benería, L. and Stimpson, C. (eds) (1987) *Women, Households, and the Economy*, New Brunswick, NJ: Rutgers University Press.

Binion, G. (1995) 'Human rights: a feminist perspective', *Human Rights Quarterly* 17: 509–26.

Brydon, L. and Chant, S. (1989) *Women in the Third World: Gender Issues in Rural and Urban Areas*, New Brunswick, NJ: Rutgers University Press.

Bunch, C. (1990) 'Women's rights as human rights: toward a re-vision of human rights', *Human Rights Quarterly* 12: 486–98.

Children's Policy Institute of West Virginia (1995) 'Executive summary of welfare to work: Breaking the cycle of poverty', Charleston, WV: Children's Policy Institute of WV.

Dwyer, D. and Bruce, J. (eds) (1988) *A Home Divided: Women and Income in the Third World*, Stanford, CA: Stanford University Press.

Fisher, S. L. (ed.) (1993) *Fighting Back in Appalachia: Traditions of Resistance and Change*, Philadelphia, PA: Temple University Press.

Folbre, N. (1986) 'Hearts and spades: paradigms of household economics', *World Development* 14(2): 245–55.

——(1995) 'Engendering economics: new perspectives on women, work, and demographic change', paper prepared for the World Bank's Annual Bank Conference on Development Economics, Washington, DC, May 1–2, 1995.

Gaventa, J., Smith, B. E. and Willingham, A. (eds) (1990) *Communities in Economic Crisis: Appalachia and the South*, Philadelphia, PA: Temple University Press.

Greene, J. W. (1990) 'Strategies for survival: women's work in the southern West Virginia coalcamps', *West Virginia History* 49: 37–54.

Hannah, K. (ed.) (1995) *West Virginia Women in Perspective*, Charleston, WV: West Virginia Women's Commission.

Hensley, F. S. (1990) 'Women in the industrial work force in West Viginia, 1880–1945', *West Virginia History* 49: 115–24.

Lewis, R. (1993) 'Appalachian restructuring in historical perspective: coal, culture and social change in West Virginia', *Urban Studies* 30(2): 299–308.

Maggard, S. W. (1994) 'From farm to coal camp to back office and McDonald's: living in the midst of Appalachia's latest transformation', *Journal of the Appalachian Studies Association*, 6: 14–38.

Mohanty, C. T. (1991) 'Cartographies of struggle: third world women and the politics of feminism', in C. T. Mohanty, A. Russo and L. Torres (eds) *Third World Women and the Politics of Feminism*, Bloomington, IN: Indiana University Press.

Mohanty, C. T., Russo, A. and Torres, L. (eds) (1991) *Third World Women and the Politics of Feminism*, Bloomington, IN: Indiana University Press.

Morrisey, M. (1989) *Slave Women in the New World*, Lawrence, KS: University of Kansas Press.

Moser, C. O. N. (1993) *Gender Planning and Development: Theory, Practice and Training*, London and New York: Routledge.

National Center for Health Statistics, Centers for Disease Control and Prevention, Public Health Service (1993) *Monthly Vital Statistics Report* 41(9).

Oberhauser, A. M. (1995a) 'Gender and household economic strategies in rural Appalachia', *Gender, Place and Culture* 2(1): 51–70.

—— (1995b) 'Towards a gendered regional geography: women and work in rural Appalachia', *Growth and Change* 26(2): 217–44.

Oberhauser, A. M., Waugh, L. J. and Weiss, C. (1996) 'Gender analysis and economic development in West Virginia', *The West Virginia Public Affairs Reporter* 13(2): 2–13.

Peters, J. and Wolper, A. (eds) (1995) *Women's Rights, Human Rights: International Feminist Perspectives*, London, New York: Routledge.

Poersch, N., Adams, G. and Sandfort, J. (1994) *Child Care and Development: Key Facts.* Washington, DC: Children's Defense Fund.

Pudup, M. B. (1990) 'Women's work in the West Virginia economy', *West Virginia History* 49: 7–20.

Pudup, M. B., Billings, D. B. and Waller, A. L. (eds) (1995) *Appalachia in the Making: The Mountain South in the Nineteenth Century*, Chapel Hill, NC: University of North Carolina Press.

Rogers, B. (1980) *The Domestication of Women: Discrimination in Developing Societies*, London: Tavistock Publications.

Seitz, V. R. (1995) *Women, Development, and Communities for Empowerment in Appalachia*, Albany, NY: SUNY Press.

Smith, B. E. (1995) 'Crossing the great divides: race, class, and gender in southern women's organizing, 1979–1991', *Gender & Society* 9(6): 680–96.

Smith, J. and Wallerstein, I. (eds) (1984) *Creating and Transforming Households: The Constraints of the World Economy*, Cambridge: Cambridge University Press.

Thomas, D. Q. and Beasley, M. E. (1993) 'Domestic violence as a human rights issue', *Human Rights Quarterly* 15: 36–62.

Tinker, I. (ed.) (1990) *Persistent Inequalities: Women and World Development*, New York: Oxford University Press.

United Nations (1995) *The World's Women 1995: Trends and Statistics*, New York: United Nations.

US Bureau of the Census (1990 and 1993) *General Population Characteristics: United States*, CP-1-1, Washington, DC: US Government Printing Office.

—— (1990 and 1993) *General Population Characteristics: West Virginia*, CP-1-50, Washington, DC: US Government Printing Office.

US Bureau of Labor Statistics (1990) *Employment and Earnings, States and Areas*, Washington, DC: US Government Printing Office.

Walls, D. S. (1976) 'Central Appalachia: a peripheral region within an advanced capitalist state', *Journal of Sociology and Social Welfare* 4: 232–47.

Weiss, C. (1990) 'Organizing women for local economic development', in J. Gaventa, B. E. Smith and A. Willingham (eds) *Communities in Economic Crisis: Appalachia and the South*, Philadelphia, PA: Temple University Press.

Weller, J. E. (1965) *Yesterday's People: Life in Contemporary Appalachia*, Lexington, KY: University of Kentucky Press.

West Virginia Bureau of Employment Programs (1993) *West Virginia Women in the Workforce*, Charleston, WV: West Virginia Bureau of Employment Programs.

West Virginia Coalition Against Domestic Violence (WVCADV) (1996) *Toward 2000: A Statewide Plan 1996–2000*, Sutton, WV: WV Coalition Against Domestic Violence.

World Bank (1995) *Toward Gender Equality: The Role of Public Policy*, Washington, DC: World Bank.

7

GENDER, INFORMAL EMPLOYMENT AND THE RIGHT TO PRODUCTIVE RESOURCES

The human rights implications of micro-enterprise development in Peru

Maureen Hays-Mitchell

INTRODUCTION

Today, nearly half a century after the United Nations' Universal Declaration of Human Rights proclaimed that 'All human beings are born free and equal in dignity and rights', the freedom, equality, dignity and rights of much of the world's population continues to be compromised by law, custom or deed. Whereas the human rights of multiple and diverse groups of individuals throughout the world are routinely violated, the rights of no group is as comprehensively affected as those of women (Human Rights Watch 1995). While women throughout the South and parts of the North experience many of the same violations of civil and political liberties as their male counterparts, they suffer additional and distinct violations strictly on the basis of their gender. The 'double injury'[1] wrought by economic restructuring and sustained by women throughout the South vividly portrays this phenomenon. Women suffer doubly, as citizens of the South and as women, the burden of structural adjustment imposed in the name of development. Increasingly in recent years, scholarly discourse has called attention to the negative effect of structural adjustment on women, focusing on the crises of survival they face and the household strategies they have devised to mitigate their deteriorating status. While this literature has contributed significantly to understandings of the differential impact of economic globalization and capitalist development within the developing world, scant attention has been directed toward the relationship between these processes and the human rights of the populations under investigation.

This chapter seeks to advance our understanding of gender-based discrimination inherent in the neo-liberal model of development currently pursued by states throughout Latin America and its relevance to the human rights community. To this end, I analyse the ways in which the structural adjustment program advocated by the government of Peru (at the behest of the international financial community) affects the economic, social and cultural rights of women in that country by examining the differential experience of economic restructuring for women and men informal workers. To begin, I examine the human rights dimension of economic restructuring by exploring the relationship between structural adjustment and the feminization of poverty as well as the influence of cultural and social norms in determining women's access to productive resources and, hence, gainful livelihoods. Next, I examine gender-specific responses to the disenfranchising conditions of restructuring by focusing on the grassroots initiatives of women informal workers to overcome their exclusion from conventional sources of finance capital and technical training.

The argument presented is twofold. First, I argue that (1) the widespread exclusion of women informal workers from conventional programs for micro-enterprise development is the result of gender biases deeply embedded in the neo-liberal model of development and upheld by patriarchal interests, ideologies and institutions within Peru, and (2) by negatively affecting their ability to access productive resources and to participate fully in the process of development, this exclusion violates the basic human right of women to development as set forth in the 1986 United Nations Declaration of the Right to Development.[2] Second, I argue that (1) the collective action of women informal workers to form or contribute to the design and operation of their own micro-enterprise development programs is a culturally and historically rooted response to this infringement of their right to development, and (2) those programs that allow, indeed expect, participating women to contribute directly to their design and operation open a new space for women to nurture the skills and strategies necessary to challenge their exclusion from the benefits of human rights and development, and that this may ultimately lead to their complete economic, political and social empowerment.

This discussion is based on fieldwork conducted in 1993 and 1994 primarily in shanty towns surrounding Lima and is the outgrowth of a national-level investigation of the informal sector in Peru conducted in 1986 and 1987. For the present study, in-depth interviews were conducted with male and female participants in a range of gender-blind and gender-focused micro-enterprise development programs and with their family members; with the directors, administrators and field personnel of selected programs; as well as with officials in private banks, parastatal organizations, government ministries and international organizations concerned with micro-enterprise development in Peru. Fieldwork also involved reviewing pertinent government legislation and program documentation in addition to accompanying selected program personnel as they conducted field visits to participating individuals and collectivities as well as accompanying a selection of women micro-entrepreneurs as they conducted their daily activities.

STRUCTURAL ADJUSTMENT, MICRO-ENTERPRISE DEVELOPMENT AND THE FEMINIZATION OF POVERTY

The present course of development in Peru is dictated by the strictures of a stabilization and structural adjustment package imposed by the international financial community to address the effect of escalating internal and external debt accumulated through years of fiscal mismanagement. Like most developing countries, Peru entered the 1980s in a state of extreme indebtedness, having borrowed freely in preceding decades to finance industrialization and the provision of infrastructure. In the context of global recession and deteriorating economic conditions, President Alan García attempted in 1985 to dictate the terms of Peru's external debt. Severed from the international financial community, economic and political conditions in Peru spiralled downward. As 1990 unfolded, GDP was found to have declined 12.5 per cent in the preceding year alone; inflation was approachng 4,000 per cent, average real wages had plummeted by nearly 50 per cent (IDB 1990: 171–2), underemployment in most urban centers was surpassing 50 per cent (MTPS 1990), and political violence was escalating.

Determined to re-establish ties with the international financial community, the government of President Alberto Fujimori agreed in August 1990 to the terms of a neo-liberal stabilization and structural adjustment program. The principal goal of the program was to stabilize the Peruvian economy by improving key macroeconomic variables (e.g. inflation, international reserves, deficit, monetary emission). A series of structural reforms designed to deregulate and modernize the Peruvian economy was set in motion. It included policies to stabilize inflation and international reserves, to liberalize markets, to privatize public enterprises, and to promote foreign direct investment. The program was further designed to reform state and related institutions. This involved reducing the size of government and of public expenditures in health, education, housing and social services as well as institutional reform. While Fujimori's stabilization and adjustment program can claim numerous achievements, especially when assessed in terms of macroeconomic variables, it must be faulted for lacking a social component that would mitigate its negative impact on the most vulnerable sectors of the population, such as low-income women and their dependent relatives.[3]

A growing body of research indicates that structural adjustment negatively affects the lives of women in gender-specific and multidimensional ways. Cook's review of the literature on women and structural adjustment identifies five ways in which women are disproportionately disadvantaged under structural adjustment programs (Cook 1994: 14). First, because social expenditures under structural adjustment are reduced rather than increased to offset recessionary-related pressures, the total burden of women's work increases. Second, employment creation is weak under structural adjustment, especially in sectors in which women predominate. Third, structural adjustment has done little to address institutional gender inequality in both the formal and informal sectors of national economies. Fourth, the tendency of neo-liberal development schemes to concentrate resources on export crops, which has traditionally worked against women in agri-culture, has intensified under structural adjustment. And fifth, because women in industry tend to operate relatively small enterprises (i.e. micro-enterprises), they experience discrimination because of the bias against small-scale operators inher-ent in structural adjustment policies. The experience of Peruvian women, espe-cially within the popular sector, corroborates this assessment. Under conditions of economic crisis and structural adjustment, most women in Peru have seen their purchasing power as food providers deteriorate as adjustment policies have induced falling wages, the elimination of food subsidies and rising prices. Cut-backs in public expenditures in health care and education have led to diminished care and training for poor women and their families while increasing their burden as the primary health-care providers and educators within families.[4] Con-sequently, most have been forced to balance greater amounts of wage work with higher levels of subsistence, domestic and community production in meeting household needs.

Research on Lima's labor market reveals that male and female participation within the paid workforce (formal and informal) has intensified concurrently with structural adjustment. It further reveals, however, that women's participation has increased at a more rapid rate than men's and that it is characterized by employment relations and work conditions that place them at the bottom of the occupational hierarchy (Scott 1991).[5] Women predominate in low-wage jobs or extremely small-scale, owner-operated ventures; these gender-segregated and poorly remunerated occupations include streetvending, domestic service, indus-trial homeworking, food preparation and repetitious manual production. Typically, women's position within the labor market is characterized by limited and insecure employment opportunities and marked by substandard wages, poor work conditions, unstable hours and disadvantageous (if any) contracts. They are further handicapped by the lack of affordable and adequate childcare as well as gender stereotyping, which denies them the requisite skills, training and capital to secure more lucrative or secure employment.[6]

It has been argued that the patriarchal ideologies and structures that maintain a sexual division of labor within the Peruvian household are also responsible for the subordination of women within the labor market (Faulkner and Lawson 1991). Indeed, focusing on gender as an axis of labor segmentation suggests a causal relationship between the patriarchal system of social relations and women's occu-pational subordination, which helps explain the persistence of gender as a dif-ferentiating factor within the Peruvian workforce (Scott 1991). Accordingly, it is alleged that patriarchal gender relations within the household directly restrict women's availability for wage work and, by influencing the allocation of resources (e.g. education or training, time and capital), indirectly condition the terms of employment for those women who do enter the workforce. In short, only occupa-tions with limited access to resources, opportunity and autonomy are accessible to many women. As a result, popular-sector women in Peru tend to be concentrated in temporary, low-paid and low-status jobs that typically replicate their household tasks. The predominance of women in gender-specific and lower-echelon occupations not only reinforces societal views of women's 'natural abilities' and 'proper place' but justifies the exclusion of women from opportunities to acquire the skills and resources necessary to compete effectively in the labor market.

As many popular-sector women in Peru turn to their own resourcefulness and attempt to produce or market goods within the informal sector, they unwittingly join the expanding ranks of Peru's 'micro-entrepreneurs'.[7] Limited capital assets combined with an ambiguous legal status exclude most informal workers, or micro-entrepreneurs,[8] from official channels of institutional credit, so some credit extension programs have been specially designed to support small-scale production and commerce. They are sponsored primarily by non-governmental organizations (NGOs) and have increasingly sought out micro-entrepreneurs in order to offer them loans at the official financial market rate. Responding to a highly visible and compelling need, such organizations have proliferated as Peru's institutional crisis has intensified.[9] Although most micro-enterprise development programs boast a gender-blind client-selection process, their philosophies and methodologies betray an inherent conceptual bias which sustains the widespread exclusion of women.

In an environment of diminishing financial support and heightened demand for accountability on the part of donor organizations, most credit-extension programs are pressed to target 'credit-worthy' clients: those whose operations are deemed most likely to respond quickly and positively to an infusion of capital as measured in terms of capital accumulation and employment creation. Accordingly, most programs seek out micro-entrepreneurs who are in relatively lucrative lines of work, have previous business experience, employ workers other than the owner and his or her family, and demonstrate the potential to expand and become profitable. Although women comprise approximately 40 per cent of the informal sector in Peru (Webb and Fernández-Baca 1994: 674), these criteria exclude most women micro-entrepreneurs. Women predominate in the least capitalized and most poorly remunerated sectors of the economy and are more likely than men either to be new entrants into the workforce or to have limited formal work experience, as well as to be working alone or with the assistance of family members.

In evaluating the 'worthiness' of the types of micro-enterprises in which women predominate, administrators of micro-enterprise development programs in Peru and elsewhere commonly dismiss such businesses as insignificant, assuming that women who operate small-scale (often home-based) businesses are not serious entrepreneurs, that the income they generate is negligible, and that their businesses hold little potential for growth. They further assume that women use financial services only for short-term consumption and do not save with an eye to the future. Data gathered from micro-enterprise development programs throughout the world indicate, however, that such presuppositions are unfounded and that women micro-entrepreneurs are indeed 'credit-worthy' (Arias 1985; Blumberg 1995; Clark 1991; Weidemann 1992; Youssef 1995). Instead, they suggest that the criteria used to determine 'credit-worthiness' fail to consider the social constraints that popular-sector women face in the circumstances of their lives and, hence, render invalid any comparison with their male counterparts. Moreover, these data suggest that the failure to consider women's lived reality reflects a gender bias deeply embedded in international development thinking and practice that works to the advantage of men and disadvantage of women.[10] Unless women micro-entrepreneurs themselves or the economic activities in which they predominate are specifically targeted by micro-enterprise development programs, women in Peru are all but systematically excluded from the benefits of these programs.

The exclusion of women from opportunities to acquire the skills and resources necessary to compete effectively in the labor market leads women, as Youssef (1995) suggests, to invest their time and resources inexpediently, which reinforces the misperception that investing in women micro-entrepreneurs is not cost-effective. When pressed to explain why their programs neglect the types of activities in which women predominate, numerous administrators of intermediate NGOs working in micro-enterprise development in Peru responded to the effect that, 'We are interested in business development, not poverty relief.' This attitude reveals that gender-based discrimination operates against women in micro-enterprise development practices. Not only does this further institutionalize women's economic dependence on (and social subordination to) men, accelerate the feminization of poverty, and exacerbate the general economic and social deterioration underway throughout Peru, but, because it directly limits the ability of women informal workers to access productive resources and skills, it constitutes a violation of their basic human right to participate fully and equally in the process of development. Hence, the experience of women informal workers in Peru appears to substantiate the recommendation that the notion of human rights 'be rethought to accommodate women's experience of disproportionate disadvantage under structural adjustment programs' (Cook 1994: 14).[11]

EMPOWERING WOMEN THROUGH COLLECTIVE ORGANIZATION

Throughout history, women in Peru – as throughout Latin America – have devised multiple, complex and innovative responses to the varying crises in their lives. Traditionally, they have relied on both formal and informal networks of mutual aid that have served to make demands on state authorities, stretch family income, extend family and neighbourhood networks of mutual assistance, and resolve community problems. In short, they have offset the debilitating effects resulting from the violation of political, civil, social or economic human rights.[12] A small group of development agencies has tapped into this heritage of collective action by directing their programs specifically toward popular-sector women. They are aware of the vital role women play in the economic and social life of countries such as Peru, the significant contribution women could make to national development if provided the opportunity, and international data (Arias 1985;

Blumberg 1995; Clark 1991; Weidemann 1992) indicating that women are a highly credit-worthy clientele.

These programs comprise a cluster of innovative methodologies which, unlike conventional (i.e. gender-biased) micro-enterprise development programs, provide credit and training for sectors in which women predominate, such as commerce and services, as well as for operations with very few employees (e.g. home-based and ambulant activities). In order to ensure that their projects address the everyday lives and needs of participating women, most agencies either seek out existing collectivities of women (e.g. *clubes de madres* ('mothers clubs'), *comedores populares* ('soup kitchens') or require that women first form a collectivity (e.g. credit cooperative, 'solidarity group') on their own. In the course of field-work, case studies were conducted of six micro-enterprise development programs operating in the *pueblos jóvenes* (shanty towns) surrounding Lima. Of these, five target women micro-entrepreneurs in particular and one targets small-scale business sectors in which women predominate (e.g. commerce and services). Of the five gender-focused programs, three utilize a variation of the 'village-banking' methodology pioneered in the early 1980s by FINCA-International and two apply a gendered methodology exclusive to their organizations. The village-banking-style programs serve collectivities of women who have come together to form credit cooperatives which, in turn, serve as a forum for gender-consciousness raising. They tend to target or attract women working in informal commerce (e.g. vendors of unprocessed or prepared foods, factory-made clothing, manufactured products, and services such as cellular-phone usage from fixed stands often in regional markets) as well as women operating small manufacturing or service businesses within their homes (e.g. assembly, repair or sale of clothing and hairstyling). The two independent programs deal with participating women on both an independent and collective basis; credit and business training are customized for each woman while training of a more social nature, but which improves the ability of women to manage their businesses, occurs when participating women and, in many cases, their families are gathered together. One tends to work with women who process and prepare food for sale within their homes or in a community center; the other tends to deal with women interested in operating non-traditional industries from their homes (e.g. carpentry, shoe-making, metal-working). Integral to the five gender-focused programs is the expectation that participating women will contribute directly to the design and implementation of the program. The gender-blind, but women-dominated, program utilizes the 'solidarity group' methodology and is devoid of a gendered dimension. It tends to appeal to women working in informal commerce (e.g. vendors of fresh produce, meats and poultry; merchants selling factory-made clothing, prepared foods or other products from fixed stands).

The women participating in the programs are known as *socias*. Interviews with them indicate that for many, even experienced micro-entrepreneurs, these programs provide the first opportunity to access working capital on non-exploitative terms as well as a rare opportunity for them to come together to

117

achieve something for themselves alone and independently of their families.[13] For a woman who has never been entrusted with credit or has never earned a cash income (clear of crippling debt obligations) on her own, this can be an empowering experience. For example, various women suggest that participating in a micro-enterprise development program has imparted to them the resources and skills to enter the labor market for the first time, to leave unfavorable employment arrangements, to develop existing businesses, as well as to switch to more lucrative lines of work. That is, they have acquired greater control over their own labor and, hence, lives.

The particular experiences of those participating in the five gender-focused programs suggest that the gendered dimension of these programs not only has led to perceptible changes in business practices and attitudes but has opened a new space for participating women to nurture the skills that ultimately may lead to their more complete empowerment. By infusing a human dimension into the technical aspects of the credit process, gender-focused programs aim to strengthen the creative capacity, initiative and confidence of participating women while providing critical information, experience and expertise. Methodologies stress the self-worth (*auto-valorización*) of individual *socias* and the need to be active agents influencing their own destinies. Recognizing the resistance that women may encounter from male partners and co-workers, gender-focused programs commonly include male family and community members in training sessions which, in turn, address the close relationship between family and community well-being and women's empowerment. As women spoke of the changes under way in their lives, it became clear that these changes are multidimensional and can be classified into four interrelated and overlapping categories: (1) physical well-being of individual women and their families, (2) women's self-perception, (3) collective consciousness of women, and (4) community welfare.

(1) Physical well-being of individual women and their families   Personal interviews with women micro-entrepreneurs participating in gender-focused programs (i.e. *socias*) indicate that nearly all have been able to diminish debt, expand savings, improve family nutrition, reduce production and marketing costs, or increase their ability to make short- and long-term investments in both their professional and personal lives (e.g. to pursue education, build a home, upgrade living or work conditions). Women who are co-heads of households indicate that their increased earning capacity contributes significantly to household incomes and is invested immediately in improved nutrition, clothing and education for children as well as upgraded housing conditions for the family as a whole. Any surplus capital is usually invested in the women's businesses and occasionally, in the form of loans, in their husband's businesses. According to women who are single heads of households, their increased income-generating capacity has proved critical to household survival and is invested principally in resources relating to their own and their children's health and nutrition.

(2) Women's self-perception   *Socias* further suggest that their personal lives are affected on another plane by participating in the programs. They claim to exert greater control over their lives in decisions concerning fertility and income expenditure and to be better able to bargain and bring pressure to bear both at the workplace and within the household. Moreover, they indicate that they have acquired a clearer sense of their rights as citizens and that they have gained respect, power and authority at home and in the community. Most *socias* attribute these changes to an enhanced sense of economic and social security due to developing their businesses, and to building solidarity with other women participating in the program. They acknowledge that such changes help them to negotiate away the resistance and opposition that they encounter, both at home and the workplace, as a result of their new status. Nearly all assert that, despite the risks involved in their new roles, they will not 'turn back'.

(3) Women's collectivities   Change is also under way at the collective level as *socias* claim that they can accomplish more for themselves, their families and communities by working together and supporting one another. Not only do *socias* prove responsible in the management of their businesses by maintaining a near-perfect repayment rate, but they convey a sense of commitment to the collectivity which is rooted in the trust that their peers have placed in them. The women know that their ability to secure future loans and training and the ability of other women to do so depends on their ability to fulfill their obligations to their collective or sponsoring organization. They also indicate that organizing collectively provides an opportunity to advise colleagues, converse with acquaintances and contribute to community-wide gatherings. Their conversations further reveal that their collectivity is a mutual support system, providing *socias* the chance to establish relationships outside the family, to recognize that their problems extend beyond their households, to vent concerns and to seek solutions collectively.

(4) Community welfare   The participation of popular-sector women in programs for micro-enterprise development also affects the communities in which they live and work. First, the programs are instruments for economic development and capital formation among *socias* which affect not only local economies but, in many cases, the physical appearance of communities: homes are upgraded, businesses established in new locales and community facilities are improved. Second, the programs can be platforms for diverse community-based activities and, occasionally, for more far-reaching projects that often bear longer-term economic benefits for communities, since development resources are commonly channeled through established organizations. Third, *socias* claim that their experience, particularly in gender-focused programs, has given them the confidence not only to suggest change but to assume the leadership in campaigns to improve the quality of life in their communities. By offering an alternative perspective, the 'gendering' of community leadership has the potential to influence the content, method and outcome of community affairs.

The type of change that has been set in motion in the lives of women participating in the case-study programs, as well as in the lives of their families and
communities, highlights the importance of providing women equal access to
development resources. It further highlights the value of assigning women significant responsibility for, and authority within, the programs in which they participate. Moreover, it suggests that, if micro-enterprise development programs are
committed to advancing development by means of empowering women, they are
more likely to succeed if their design and content accept that the needs, interests
and capabilities of women informal workers in Peru are, in many regards, unique,
far-reaching and gender-specific. Finally, the experiences, perspectives and
insights conveyed by the case-study women micro-entrepreneurs – both in terms
of their exclusion from conventional programs and their empowerment through
gender-focused programs – reveal to us that the 'double injury' of structural
adjustment violates the basic human right to development not only of women but
of the entirety of Peruvian society. Likewise, they reveal that efforts to ameliorate
the negative effects of structural adjustment on women, such as those discussed
here, hold the potential to ameliorate its negative impact on Peruvian society as a
whole.

CONCLUSION

'Gendering' the link between development and human rights praxis reveals the
relevance of broadening the concept of 'human rights' to embrace the everyday
experiences and perspectives of popular-sector women. Accordingly, the case
study presented here illustrates how the exclusion of women from programs for
micro-enterprise development in Peru limits their ability to access productive
resources and, thus, to participate equally in development. In doing so, it violates
the basic right of women to development as set forth in the 1986 United Nations'
Declaration of the Right to Development, which ensures the right of all individuals, regardless of gender and geography, to participate in and benefit from
socio-economic development. Moreover, it uncovers the structural discrimination
on which economic restructuring in Peru and, by extension, the prevailing model
of development within Latin America is based.

The experiences of women micro-entrepreneurs in Peru suggest that simply
including women in programs of micro-enterprise development will not be sufficient to overcome the 'double injury' they sustain from structural adjustment.
But rather, the premiss on which neo-liberal restructuring operates should be
reconsidered, in order to take into account non-economic factors that contribute
to the gender-specific experience of structural adjustment in developing societies
such as Peru. Throughout the course of fieldwork, women micro-entrepreneurs,
and in many cases their family members, emphasized their deep conviction that,
by participating in micro-enterprise development progams (especially those that
are gender-focused), they are acquiring a space from which to challenge the
disenfranchising effects of economic crisis and restructuring. However, if this is

the case, the harsh reality of their everyday lives makes clear that their new 'space' is precarious. In and of itself, micro-enterprise development represents only a preliminary step toward the more complete empowerment of women within Peruvian society. Before embracing it, we must be mindful that, at present, too few women have the opportunity to participate in such programs and that too few programs are gender-focused in nature. Moreover, it is too easy for state and international actors to relinquish their responsibility for ensuring the socio-economic rights of women to the grassroots initiatives of popular-sector women and the local agents willing to work with them. Before final determination can be made, analysis of the long-term impact of, and problems associated with, all types of micro-enterprise development is needed.

In order to transcend the artificial dichotomy which privileges male notions of human rights and disregards the experiences that daily threaten women's basic rights, we must understand that most gender-based discrimination and abuse is, in the words of Charlotte Bunch (1995: 14), 'part of a larger socio-economic web that entraps women, making them vulnerable to abuses that cannot be delineated as exclusively political'. Moving women from the margins to the center of human rights protection involves more than simply grafting consideration of gender-based abuses on to the concept of human rights. It involves questioning the most basic concepts of our social order so that women's rights are understood as fundamental human rights (Bunch 1995; Friedman 1995; Peters and Wolper 1995). In the context of development rights, this means analyzing non-economic dimensions of male control over women, identifying the changes in gender rela-tions and social systems that are necessary to include the voices, experiences, insights and perspectives of women (and other groups) who have been excluded from the benefits of human rights and development, and analyzing the specific historical and cultural settings within which disenfranchised peoples of Latin America live and work. Only then can the politically, economically and socially disenfranchising conditions of neo-liberal restructuring be supplanted by more sensitive and transformative approaches to the development and human rights needs of all social constituencies.

NOTES

1 For elaboration of this concept, see Kuenyehia 1994.
2 The 1986 United Nations Declaration of the Right to Development ensures the right of all individuals, regardless of gender and geography, to participate in and benefit from socio-economic development.
3 For a thorough analysis of stabilization and structural adjustment in Peru, see Gonzales de Olarte 1996.
4 See Tanski 1994 for a more detailed discussion of the impact of structural adjustment on women in Peru.
5 Throughout the economic and political crisis in Peru, women's labor market participa-tion has increased, as a percentage of the total economically active and employed population, from 24% in 1981 (FLACSO 1993: 46) to 38% in 1993 (Webb and Fernández-Baca, 1994: 674). Within the informal sector, women's employment has

increased as a proportion of their total employment, climbing from 35% in 1981 (FLACSO 1993: 52) to 52% in 1993 (Webb and Fernández-Baca 1994: 676) and is assumed to be yet higher in the late 1990s.

6 See Berger and Buvinic 1989; Hays-Mitchell (n.d.); and Weidemann 1992 for discussion of the relationship between gendered identities and women's position within the urban labor market.

7 Though definitions vary, one widely accepted assessment is that a micro-enterprise in Peru employs fewer than four workers and has a working capital not exceeding US$300 per employee (Webb and Fernández-Baca 1994: 671). It is estimated that the number of micro-enterprises operating in Peru increased from 75,000 in 1988 to over 100,000 in 1994 and is projected to reach approximately 135,000 by 2000 (ibid.: 671). Of these, approximately 40 per cent are believed to be operated by women and to fall far short of the maximum working capital cited (ibid.: 674).

8 Though not synonymous, terms such as 'micro-entrepreneurs' and 'micro-enterprise sector' are displacing terms such as *informales* and 'informal sector' in Latin American development discourse. It is not uncommon to find the terms 'micro-enterprise' and 'informal' as well as statistics relating to them used interchangeably. The rise in popularity of 'micro-enterprise'-related terms is rooted in two motives. First, it focuses attention away from the legality of an operation (e.g. as registered or tax-paying) and away from the false dichotomization of developing economies and, instead, toward the questions of scale and wealth. Second, it is more consistent with neo-liberal discourses extolling entrepreneurship and the virtues of the free market.

9 In 1993, it was estimated that nearly 500 NGOs operated in Peru and, of these, approximately 120 were directly involved in micro-enterprise development (APOYO 1993; PACT 1993).

10 For more complete discussion of the notion of gender bias in development, see Elson 1991a.

11 This recommendation was formally advanced by a consultation of international lawyers representing women's interests in Africa, the Americas, Asia, Australia and Europe held at the Faculty of Law of the University of Toronto in August 1992 and reported in Cook 1994.

12 See Arizpe 1990; Bourque 1989a, 1989b; Jaquette 1991; Jelin 1990; Logan 1990; Nash and Safa 1986; Safa 1990 and Vargas 1992 for elaboration on the heritage of collective organization among Latin American women.

13 For discussion of the ways in which the identities and interests of popular-sector women are subsumed by those of their families, thereby hindering them from forming well-defined notions of what they as individuals want and need, see Elson 1991b; Hays-Mitchell (1999); and Moser 1993.

REFERENCES

APOYO (1993) 'Estudio de impacto de las ONGs en la lucha contra la pobreza en el Perú', Lima: APOYO, unpublished document.

Arias, M. (1985) 'Perú: Banco Industrial del Perú: credit for the development of rural enterprise', in C. Overholt, M. Anderson, K. Cloud and J. Austin (eds) *Gender Roles in Development Projects: A Case Book*, Boulder, CO: Westview Press.

Arizpe, L. (1990) 'Democracy for a small two-gender planet', in E. Jelin (ed.) *Women and Social Change in Latin America*, London: Zed Books.

Berger, M. and Buvinic, M. (1989) *Women's Ventures: Assistance to the Informal Sector in Latin America*, West Hartford, CT: Kumarian Press.

Blumberg, R. (1995) 'Gender, microenterprise, performance, and power: case studies from the Dominican Republic, Ecuador, Guatemala and Swaziland', in E. Acosta-Belén and

C. Bose (eds) *Women in the Latin American Development Process*, Philadelphia, PA: Temple University Press.

Bourque, S. (1989a) 'Gender and the state: perspectives from Latin America', in S. Charlton, J. Everett and K. Staudt (eds) *Women, the State, and Development*, Albany, NY: State University of New York Press.

—— (1989b) 'Urban development', in K. Stoner (ed.) *Latinas of the Americas*, New York: Garland Press.

Bunch, C. (1995) 'Transforming human rights from a feminist perspective', in J. Peters and A. Wolper (eds) *Women's Rights, Human Rights: International Feminist Perspectives*, New York: Routledge.

Clark, J. (1991) *Democratizing Development: The Role of Voluntary Organizations*, West Hartford, CT: Kumarian Press.

Cook, R. (1994) *Human Rights of Women: National and International Perspectives*, Philadelphia, PA: University of Pennsylvania Press.

Elson, D. (ed.) (1991a) *Male Bias in the Development Process*, New York: St Martin's Press.

—— (1991b) 'Male bias in the development process: an overview', in D. Elson (ed.) *Male Bias in the Development Process*, New York: St Martin's Press.

Faulkner, A. and Lawson, V. (1991) 'Employment versus empowerment: a case study of the nature of women's work in Ecuador', *Journal of Development Studies* 27(4): 16–47.

FLACSO (1993) *Mujeres en Cifras*, Madrid: Ministerio de Asuntos Sociales de España and Santiago: Facultad Latinoamericana de Ciencias Sociales.

Friedman, E. (1995) 'Women's human rights: the emergence of a movement' in J. Peters and A. Wolper (eds) *Women's Rights, Human Rights: International Feminist Perspectives*, New York: Routledge.

Gonzales de Olarte, E. (ed.) (1996) *The Peruvian Economy and Structural Adjustment: Past, Present and Future*, Miami, FA: North-South Center Press.

Hays-Mitchell, M. (1999) 'From survivor to entrepreneur: gendered dimensions of micro-enterprise development in Peru', *Environment and Planning A: The International Journal of Urban and Regional Research* 30 (forthcoming).

Human Rights Watch (1995) *The Human Rights Watch Global Report on Women's Human Rights*, New York: Human Rights Watch.

IDB (Inter-American Development Bank) (1990) *Economic and Social Progress in Latin America*, Washington, DC: IDB.

Jaquette, J. (1991) 'Introduction', in J. Jaquette (ed.) *The Women's Movement in Latin America: Feminism and the Transition to Democracy*, Boulder, CO: Westview Press.

Jelin, E. (1990) 'Citizenship and identity', in E. Jelin (ed.) *Women and Social Change in Latin America*, London: Zed Books.

Logan, K. (1990) 'Women's participation and urban protest', in A. Craig and J. Foweraker (eds) *Popular Movements and Political Change in Mexico*, Boulder, CO: Lynn Rienner Publishers.

Kuenyehia, A. (1994) 'The impact of structural adjustment programs on women's international human rights: the example of Ghana', in R. Cook (ed.) *Human Rights of Women: National and International Perspectives*, Philadelphia, PA: University of Pennsylvania Press.

MTPS (Ministerio de Trabajo y Promoción Social) (1990) *Boletín Mensual de Empleo*, Lima: Ministerio de Trabajo y Promoción Social, Dirección General de Empleo.

Moser, C. O. N. (1993) *Gender Planning and Development: Theory, Practice and Training*, London and New York: Routledge.

Nash, J. and Safa, H. (1986) *Women and Change in Latin America*, South Hadley, MA: Bergin and Garvey.

PACT (Private Agencies Working Together) (1993) 'Estado situacional de las organizaciones no gubernamentales de desarrollo en el Perú', Lima: PACT, unpublished document.

123

Peters, J. and Wolper, A. (1995) 'Introduction' in J. Peters and A. Wolper (eds) *Women's Rights, Human Rights: International Feminist Perspectives*, London and New York: Routledge.

Safa, H. (1990) 'Women's social movements in Latin America', *Gender and Society* 4(3): 354–69.

Scott, A. (1991) 'Informal sector or female sector? Gender bias in urban labour market models', in D. Elson (ed.) *Male Bias in the Development Process*, New York: St Martin's Press.

Tanski, J. (1994) 'The impact of crisis, stabilization and structural adjustment on women in Lima, Peru', *World Development* 22(11): 1627–42.

Vargas, V. (1992) *Como Cambiar El Mundo Sin Perdernos: El Movimiento de Mujeres en el Perú y América Latina*, Lima: Editorial Flora Tristan.

Webb, R. and Fernández-Baca, G. (1994) *Perú en Números 1994*, Lima: Instituto Cuánto.

Weidemann, C. (1992) *Financial Services for Women*, Bethesda, MD: Development Alternatives.

Youssef, N. (1995) 'Women's access to productive resources: the need for legal instruments to protect women's development rights', in J. Peters and A. Wolper (eds) *Women's Rights, Human Rights: International Feminist Perspectives*, New York: Routledge.

8

GENDER, MIGRANTS AND RIGHTS IN THE EUROPEAN UNION

Eleonore Kofman

INTRODUCTION

'The West, the champion of universalism and inalienability, has been reluctant to recognise the equality of migrants with nationals, and migrants' rights have been something of a Cinderella in the family of human rights' (Ghai 1997). Consideration of migrant needs and the transnationalisation of entitlements (Baubock 1993) have not kept pace with the globalisation and diversification of migration and population movements. Not only has the adoption of an International Convention for Migrant Workers and their Families (United Nations 1991), passed in 1990, been a protracted process but its impact has been more symbolic than substantive. No European state has so far ratified it. The reasons for the slow progress and vexed negotiations lie in the jealous guarding of state sovereignty and identity of states. Although the Convention does not concern the process of migration but the situation of migrants, primarily legal and regular ones, in receiving states, it touches upon the desire of these states to preserve the right to control and regulate the distinctions between citizens and different categories of non-citizens. The mute response towards the Convention underlines the reluctance of European states to acknowledge the role of migration in their past histories, their present situation or future trajectories, and has implications for equality of treatment and recourse to anti-discrimination legislation for migrants and citizens across a range of policy and planning areas, such as in education, employment, health and housing.

The failure to ratify the Convention does not mean that the rights of migrants in European states uniformly fall short of its recommendations. The boundaries of rights (Baubock 1991) between citizens and legal long-term residents (denizens) have been substantially reduced in recognition of the latter's embeddedness in European societies (Ansay 1991). Social rights, in particular, tend increasingly in welfare states to operate largely on the principle of territorial coverage of all legal residents, though the actual exercise of these rights continues to be restricted in many states through differential rights to take up employment. At the same time, sharper distinctions have been drawn between denizens and undocumented migrants and asylum-seekers in relation to civil, economic and social rights. While some convergence has taken place between member states of the European Union, especially in the control of clandestine migrants and asylum-seekers, rights of migrants still depend largely on their specific situation within each state. The European Union has, on the whole, failed to address the issue of the rights of third country nationals (migrants from non-European Union states), while dismantling the barriers between citizens of different member states.

Though international conventions mainly concern the situation of migrants after their entry into a state, the status of migrants, especially in the initial years of residence, will be strongly influenced by the way in which entry is gained. A key distinction in relation to rights is the legal situation of migrants, that is, whether they are in a regular situation or undocumented. Quite commonly, undocumented migrants enter legally as tourists or short-term workers and then overstay. It is difficult to estimate the numbers of the undocumented, except to state that they form a large proportion of the new flows. The most dramatic changes in the past decade have occurred in the extension of immigration to southern European countries, new East–West flows following the end of the Cold War and a sharp increase in refugees and asylum-seekers. The duration of these new and under-studied flows is not clear, although in the case of the East–West flows there is evidence of circulatory migration rather than long-term residence or settlement (Morokvasic and Tinguy 1993). At the same time, a steady flow of family reunification migrants, whose intention is to settle, has continued in most European Union states.

Each of these flows also has a distinctive gendered division for which unfortunately we still only have sketchy information. Although in southern Europe, independent female labour migration is common (Campani 1993, 1997; Escrivá 1997), it is less so in northern Europe. There, women have tended to a far greater extent to enter as family reunification migrants, which has been the dominant form of legal permanent migration in the past two decades (OECD 1993). In family reunification, the conditions of entry and post-entry rights operate on the assumption that the migrant is a dependant of a 'primary' migrant, who is the breadwinner and male head of household. As the dominant type of migration associated with settlement, family reunion has been subjected to scrutiny, more restrictive criteria and vigorous application of regulations. Such measures reflect a populist political response to the renewed politicisation of immigration and the expansion of immigrant and ethnic minority communities.

Yet, at the same time, family reunion has now been incorporated into the UN Convention on the Rights of All Migrant Workers and Members of their Families, which sought to attenuate the logic of market forces and view migrant workers as more than labourers, that is, as social entities with families with rights, including that of family reunification (Hune 1991: 808). Those seeking to extend the protection and rights of migrants internationally have argued that family reunion is a right derived from normative principles based on the right to family life (Baubock

1991; JCWI 1994a). In contrast, as international organisations have increasingly brought family reunification into the domain of human rights, a number of European states have in effect made the right to live in a family more discretionary for non-citizens. The European Union has begun to propose the harmonisation of criteria of access to family reunion, in 1993 drawing up guidelines based on the lowest common denominator and a narrow definition of the what constitutes the family (JCWI 1994a, 1994b; Weber 1993).[1]

Though European states and the European Union continue to deploy a discourse of rights, these have been completely overshadowed by the imperatives of immigration control, thus turning the substance of 'rights' into discretionary practices in relation to family reunion. In effect, changing places across borders (discounting citizens of the European Union) disenfranchises migrants to varying degrees. Increasingly those who stray from a fully documented status may cut themselves and their children off from economic, social and political rights lasting for generations. Regularisation programmes (in France in 1981–2; in Italy in 1986 and 1990), in which undocumented migrants could acquire recognition and legal status in the country of employment, are no longer operative. Migration for women may result in substantial loss of rights as they may be pushed during the early period of their lives in a new country into a private patriarchy that they would no longer experience in their countries of origin. Many states, as we shall see, impose probationary periods in which dissolution of marriage becomes a grounds for deportation and thus makes the person entering the country a dependant in the full sense of the term. They may therefore be forced to choose between domestic violence and deportation, and often without the benefit of support from family or friends and an independent income. In most countries today, women do not lose their rights upon marriage, at least not to this extent.

I cannot comprehensively analyse in the space of this chapter the relationship between international conventions of human rights, state entitlements and the actual experience of migrants across a range of rights. Rather I shall limit the discussion to rights of family reunification and formation and its implications for some of the crucial policy areas of employment and housing, which have proved to be the major obstacles for women even more than men. This is a particularly significant area, since family reunification and formation currently constitute the largest flow of legal migrants in Europe and is subject to more stringent controls and their vigorous application. For women, family-related migration has been by far the dominant mode of entry, whether they enter as wives and fiancées, or themselves apply to bring in family members. In each case, they encounter distinctive problems in meeting the regulations prescribed by states. However, despite the importance of this mode of entry in the post-war history of European migration, especially in northern Europe, the taken-for-granted correspondence between female and family reunification has meant there has been little interest in examining the nature of this form of immigration and its changing composition and conditions in recent years.

I shall therefore first outline the changing nature and context of immigration since the late 1970s in Europe with particular emphasis on women. In subsequent sections I shall examine the coverage of international conventions in relation to migrant rights, the experiences of women in family reunion and the ways in which changing regulations are likely to affect their ability both to enter as family migrants and to bring in family members. In the concluding section, I shall assess the failure by European feminist movements to address these issues, on the one hand, and the possibility of developing more progressive policies through political action and campaigns, on the other.

WOMEN, GENDER RELATIONS AND EUROPEAN IMMIGRATION

Before the stoppage of mass labour migration in 1973–4, women migrated to north-western European countries as independent workers, as workers with their husbands or male partners, and as part of family migration. In countries with colonial ties, and particularly the United Kingdom, the higher proportion of family migration meant that the flows were more evenly balanced in gender terms. Since the 1970s, family reunification and formation have been generally the principal means of legal entry into northern Europe except for a limited number of professional migrants. Family reunion in the narrow sense refers to primary migrants bringing in close members of their families, while family forma-tion concerns established migrants and citizens marrying non-citizens. The latter category too has grown with the expansion of second and third generations born in the state, as well as more international intermarriages between citizens and non-citizens. As a result of these developments, it would be wrong to equate family reunion and formation with women given that growing a number of fiancés and husbands are entering under this category in Britain (Home Office 1996) and France (Tribalat 1996).

In southern Europe, the demand for domestics to fill the gaps in the welfare state as more middle-class women found employment in the formal sector meant that a number of the immigrant nationalities were heavily female (Hoskyns and Orsini-Jones 1995). In these former countries of emigration a formalised system of family reunion is less well developed (Campani 1997). In Italy it is only since 1988 that foreign workers with a regular resident permit can apply for family reunion. The segmented and fragmented labour markets in which migrants work in Italy have made it more difficult to fulfil the requisite conditions. In addition, employment in the domestic sphere may make it more complicated to bring in a family member. Family reunion is particularly low amongst female-dominated groups such as the Cap Verdians and Filipinos.

East–West migration takes on different forms again. The highest rates of migration to Western Europe are from Poland, where many women in particular engage in circulatory and temporary movements, enabling them to juggle employment and household responsibilities (Morokvasic 1993). Professional and skilled migrants form a substantial proportion of Polish migrants. Many have taken unskilled and deskilled employment, resulting in what has been called brain waste (Morawska and Spohn 1997; Morokvasic and Tinguy 1993). No analysis exists, however, of the degree to which this is happening, although studies of immigration to Mediterranean countries have elucidated the high level of education amongst certain groups such as the Filipinas in Italy (Campani 1993) and Latin Americans in Spain (Escrivá 1997). Their qualifications in the health sector may be transferred to and sought after in 'unskilled' domestic work.

Another major group eventually settling in the European Union has been the growing number of asylum-seekers who have arrived throughout the 1980s and into the 1990s. Many states introduced draconian measures to dampen the flows and limit to a very small percentage the numbers granted full refugee status, which brings with it in practice a virtually unconditional right to family reunion. Owing to the lower proportion of women granted refugee status, there has been much international campaigning to recognise gender-specific reasons in the process of asylum determination. These may include rape and escape from oppressive conditions against women in general within a particular society, such as genital mutilation or pressure to dress in particular ways. They may also include fear of persecution from association with political activists as well as political activism of a less formal kind. Australia (1996), Canada (1993 revised 1996) and the USA (1995) have all passed guidelines for the handling of gender-specific persecution which are to be taken into account in claims for refugee status (Crawley 1997), but this step has not been followed by any European country. Women may also find it more difficult to acquire the resources to flee a zone of conflict or personal persecution. In the absence of statistical information, it is thought that about a third of asylum-seekers are women. Few will be granted full refugee status; a far greater percentage are given exceptional leave to remain, a temporary form of protection permitting family reunion after four years of residence permits but, crucially, requiring the person to fulfil normal conditions of stable employment and adequate housing. Furthermore, a growing proportion of asylum-seekers are being rejected, though many stay as undocumented migrants with no rights whatsoever.

INTERNATIONAL CONVENTIONS, EUROPEAN STATES AND MIGRANT RIGHTS

Early initiatives in the area of human rights did not directly concern themselves with the situation of migrants or aliens. Rather it has been the International Labour Office (ILO) which has fought for the rights of migrant workers. In 1975 it adopted Recommendation 151 Concerning Migrant Workers, in which member states were urged to take 'account not only of short-term manpower needs and resources but also of the long-term social and economic consequences of migration for migrants as well as the communities concerned' (art. 1), including for the first time a strong emphasis on the reunification of families. Only Italy, amongst European states, signed the 1975 ILO Convention. Migrants in

European states are protected by the basic human rights conventions (right to life, against torture, freedom of opinion), but more substantive rights and protection remain the prerogative of individual states (Goodwin-Gill 1989).

Most European conventions apply only to citizens of member states. The European Convention on the Legal Status of Migrant Workers, for example, was ratified by Sweden, Portugal, Spain, Turkey, the Netherlands and Norway and came into force in 1983. It is not about conferring rights but regulating agreements between parties or states (Plender 1991: 241). The one significant exception, which encompassed all residents, was the European Convention on Human Rights (in operation since 1953); this has been invoked in support of the right of family reunion amongst third country nationals. Article 8 states that 'everyone has the right to respect for his private and family life, his home and his correspondence'. The law is dynamic and recent judgements have been weighted against the right to family life of the immigrant. The European Court of Human Rights has upheld a state's right to control immigration and argued that family life may be pursued without insurmountable obstacles by members of a family living in different states (JCWI 1996: 9).

The International Convention on the Protection of the Rights of All Migrant Workers and Members of their Families (1990), considered by some as a landmark, simply represents an extension of the general regime of human rights to migrants; it has not been ratified by any European state. One advance is that for the first time there is an explicit recognition of the active involvement of women in migration (Hune 1991). It has taken several decades for the situation of women migrants to be taken seriously in international conferences and conventions. Their marginal position was clearly expressed in the single paragraph that appeared in the recommendations drawn up at the World Conference of the International Women's Year in 1975 in Mexico. By the 1985 Nairobi Meeting several new concerns had been added to those of women as workers and their families. The Convention reduced the invisibility of women migrant workers; Article 2(1) explicitly stated the principle of equal treatment. Part III has several articles which could be used to defend against the extreme vulnerability of women in areas of sexual exploitation, physical abuse and forced prostitution, though none of these has been specifically enunciated. These are issues of growing concern in Europe (Leidholdt 1996) but have so far only resulted in extremely weak measures. The Convention sought to ensure that migrant women would be less dependent and were no longer placed in jeopardy in case of death or dissolution of marriage, a matter of considerable relevance to migrant women in Europe. However, it does not address the special situation of migrant women, their sexual exploitation or the victimisation of women.

As for family reunion, neither the vexed negotiations in the 1980s surrounding the UN Convention nor the text adopted by states following the Cairo Conference on Population and Development mentioned rights or principles for the family reunion of legal migrants. Although the Convention (United Nations 1991) reiterates article 10 of the Convention on the Rights of Children (the right of families to live together) and recognises the basic importance of family reunion and the encouragement of the incorporation of family reunion for legal migrants into national legislation, it precedes it with a statement to the effect that 'implementation of the recommendations contained in the programme stems from the sovereign rights of each state in conformity with its own national laws'. It was not just European states that rejected any references to the discourse of rights; for settler societies the basic consideration is that family members must fit into the annual quotas fixed by each state. States have also guarded their right to decide the distribution of rights between citizens and different categories of non-citizens.

In fact, the Convention still allows each state to determine the general conditions of reunion, to define the constitution of the family (formation and members) and under specific conditions, and to set the resources necessary in order to bring in family members. In some countries, what may be defined as a right is in actual fact hedged in by a series of conditions, which ensures that family reunion is achievable by fewer migrants; in others, such as the United Kingdom, immigration and settlement are surrounded by a battery of discretionary administrative procedures and decisions made by local officers.

France presents an interesting case of a country in which family reunion figured in the 1945 immigration regulations and for which specific procedures were instituted. In the 1980s, the groundswell against immigrants, orchestrated by far right parties, and debates about the integration of immigrant-origin populations have helped to push immigration flows to the fore of the political agenda. As the only legal right of entry apart from refugee status, family reunion was scrutinised by the new right-wing government in 1993 and revisions to the 1945 Code were enacted (GISTI 1994b: 83). The implementation of stricter regulations (increase in the period of residence before submitting an application, withdrawal of family allowances from the calculation of the amount of income required and the necessity to bring in all members at the same time) have been highly effective in reducing the numbers who have applied for and entered under family reunion. These fell from 32,421 family members in 1993, to 20,645 in 1994 and again in 1995 to 14,360. Similarly there was a decrease in spouses of French citizens from 20,062 in 1993 to 13,387 in 1995 (Tribalat 1996: 144; 1997: 183). Total immigration accordingly dropped dramatically from 94,000 in 1993 to 50,000 in 1995.

British regulations in this area are highly discretionary, with decisions being formed by subjective considerations. The 'primary purpose rule' here has been used increasingly to target Third-World migrants on racial criteria. 'Primary purpose' refers to the fact that the main intention of marriage cannot be to gain admittance. It has been the main measure applied against men deemed to be 'using marriage as a device' to enter the labour market (Bhabha and Shutter 1994: 77). It became the principal reason, rising from 18 per cent in 1982 to 73 per cent in 1983, for refusing applications from male dependants from the Indian subcontinent (ibid.: 80). With the election of the Labour Government in May 1997, the primary purpose rule was abolished.

This rule certainly ran contrary to the recommendations of the Committee on

131

Civil Liberties and Internal Affairs of the European Parliament, which asked member states not to manage immigration policy only from the point of view of limiting entry at the borders, but also to consider foreseeable demands created by the employment market (JCWI 1994a: 24–5). It reminded states that they are bound by international conventions and that there is no valid justification for the attitude of states which, by creating financial obstacles to bringing in family members make the right of family reunion devoid of any substance. It also recommended that social rights and access to work for third country nationals be aligned with those granted to EU migrants.

Women and family reunion

Family reunion procedures generally presumed that a migrant bringing in a spouse and children was a male. Increasingly this assumption has less and less relation to the reality, not just in southern Europe with its higher proportion of independent immigrant women, but also in northern Europe where family reunion has acquired a more masculine profile. For example, in the United Kingdom, 38.9 per cent of the 32,620 spouses accepted for settlement in 1995 were male (Home Office 1996, table 6.3). In France, 16.3 per cent in 1989 and 23.8 per cent in 1993 of family reunion spouses were males (Lebon 1994: 102).

What specific problems do women encounter in family migration? It first has to be said that family reunion, and women's experiences within it, have been under-researched (Kofman 1997; 1999). French academic and policy-oriented studies have contributed more than any other sources to our knowledge of the forms of family constitution and the problems encountered in bringing in members of the family (Hu Khoa and Barou 1996; Tribalat 1996). Women partake in family reunion in a number of ways, and each form has its distinctive characteristics and difficulties. Firstly, and still the most common, are women who enter as wives or to marry male migrants, who are usually from their country of origin. Secondly, and now a growing number with the expansion of the second and third generation of settled immigrant populations, are the women who themselves apply to bring in a spouse. Thirdly, and to a lesser extent in northern than in southern Europe, are independent women labour migrants. Lastly, there is a growing number of women refugees who are becoming eligible in terms of years of residence and roots in a society to apply for family reunion. For the first group the issue is the dependency relationship into which women are forced for, in some cases up to five years, by immigration and settlement legislation. For the other groups, the most pressing concerns are likely to be the accumulation of resources (employment, income, housing) necessary to fulfil the conditions of family reunion. I shall briefly outline the major policies affecting women specifically, or more acutely than men.

Dependency

The general relationship of dependency, in which women's rights are derived from a 'protector', is a reality of women's lives and reinforces the image of passive immigrant women following the male. Although feminists have criticised images of immigrant women bereft of agency, it is difficult to dispel this view given the economic and social relationships imposed upon women in the first few years after entry. All countries impose a probationary period ranging from a year to five years, during which the spouse's status is linked to the husband and the dissolution of marriage constitutes grounds for the revocation of the residence permit. In Germany, family reunion can remain in doubt until a permanent residence status is obtained after eight years. Inevitably this forced dependence leads women into remaining in violent relationships (Essed 1996; Kofman and Sales 1992). In Britain, 755 women were threatened with deportation in the period from January 1994 to July 1995 because their marriage had broken down (Southall Black Sisters 1997). How many others have stayed in this situation?

Employment

Family reunion confirms the idea of immigrant women as exclusively wives and mothers with a marginal economic contribution. For many years, British policy allowed men to bring in wives and fiancées but not the reverse. In its response to the charges by the European Court of Human Rights of sex discrimination against women who wished to bring in husbands and fiancés, the British Government argued that this discrimination was justified since 'women as bread-winners are unusual, for society still expects the men to go out to work and the woman to stay at home . . . The majority of women do not threaten the labour market, particularly women from the Indian sub-continent' (cited by Bhabha and Shutter 1994: 73). We actually know very little about the time that elapses between the arrival of family members and their entry into the labour market (OECD 1993: 21). Restrictions on employment immediately on entry have certainly been eased. A year's waiting period was lifted in Germany in the early 1990s, but even so the issuing of a work permit still depends for the following five years on the state of the labour market. This has meant that many women had little alternative initially other than to seek work in the undeclared informal sectors. Culturalist explanations based on Islamic background, without consideration of the factors cited above, are often proffered to account for low rates of participation in the labour force, but more rigorous analysis and evidence reveals the effect of racial discrimination on women's involvement in the labour force (Harzing 1995).

For women applying to bring in family members, the most pressing problems are likely to be the accumulation of sufficient resources and stable employment. Women's earnings are lower than men's and immigrant women's in general lower than those of the majority of women (Harzing 1995; Seifert 1996: 425). Amongst a number of immigrant women, such as those from Morocco, Turkey, Pakistan and Bangladesh, temporary and part-time work was far more common than among non-immigrant women. At a time when employment has become more precarious and flexible, the requisite conditions in most countries stipulate regular employment. Those undertaking undeclared jobs will not be able to provide a formal contract. Increasingly, benefits cannot be included as part of the available resources. For example, in France the generous family allowances can no longer be included in the SMIC (minimum wage), which is the minimum required income. In the UK, the entry of a family member cannot lead to a recourse to public funds, which in reality is likely to disqualify the unemployed. British evidence (Kofman and Sales 1997) shows that few refugee women have managed to obtain reasonably paid employment despite their educational levels. Many have children with them and are unable to work or obtain training, so that they rely on public funding. In many states dependence on public funding immediately eliminates an immigrant or refugee from qualifying for family reunion.

Housing

French research has indicated that housing is seen as the main stumbling block preventing migrants from qualifying to bring in family members. Housing short-ages have become more acute with the reduction in the construction of social housing. It may well explain the slump in family reunion in the Paris region where housing is relatively expensive (Tribalat 1997: 181–2). Amongst second generation women, who are more likely to be able to count on the aid of their family, this may present less of a hurdle. Women employed in domestic labour, the most common source of employment for women immigrants in southern Europe, are frequently housed by the employer. This circumstance may be one of the main reasons for the low rate of applications from Cap Verdian and Filipina women in Italy (Campani 1997). Where women are accompanied by children, these difficulties are compounded since the required housing standard (precisely calculated in terms of national norms in France and in Germany) is raised according to the size of the family (JCWI 1994b).

Not all immigrant women work in the less well paid sectors; an increasing number of highly skilled labour migrants are women. In Britain 16.7 per cent in 1984 and 19.9 per cent in 1992 of those granted long-term work permits (mainly given to the highly skilled) were women. However, for the majority of women wishing to bring in a family member, these conditions make access to family reunion extremely difficult. To make family reunion a reality, many organisations that support the rights of immigrants have called for the withdrawal or easing of the increasingly stringent criteria of employment and housing. For those enter-ing as family reunion migrants it is imperative that the practice and idea of dependency is challenged as strongly as possible.

134

Responses from women's movements

The difficulties of fulfilling the conditions for family reunion and issues of dependency, which generate more acute problems for women, are not, as we have seen, matters dealt with in the UN Convention. Immigrant women's concerns have not really been seriously addressed by the European Union (Harzing 1995) or national women's movements. The gains from national and European Union initiatives have largely accrued to white, middle-class and educated women, cutting off the privileged from concerns about others. Women are relying more and more on other women (increasingly immigrant) to enable them to take paid employment (Hoskyns 1996a). The European Union reports on racial discrimination, still an area of policy vacuum, usually omit the specificity of restraints and restrictions faced by migrant and black women (Hoskyns 1996a: 180–1). The action taken on sex discrimination, in contrast to the non-action on racial discrimination, is quite striking. Diverging views on priorities and strategies between mainstream, feminist and anti-racist organisations have meant that solidarity between and amongst these groups on ways of giving support to migrant and black women's issues cannot be assumed. Certainly a number of European-wide studies have been conducted on behalf of the Council of Europe and the European Union (European Women's Lobby 1995; Gaspard 1994; Lutz 1994) but there is little sign that they will lead to concrete policies to improve the position of immigrant women and give them their due voice.

It is not merely at the European level that migrant and black women are weakly represented and their experiences marginalised. In many national women's movements, these issues are for the most part ignored or treated very partially (Essed 1996; Kofman 1997; Lloyd 1998), leaving migrant and black women to discuss and organise separately. The situation is far from uniform. In Britain, black and immigrant-origin feminists began in the 1980s to challenge the views of white feminists; through their presence in academic circles, they have articulated new insights concerning the interaction of gender, class and 'race' (Anthias 1993; Anthias and Yuval-Davis 1993; Brah 1993). So too have Italian feminists taken on issues pertaining to migrant women. To some extent this activity may also be due, despite their presence in the domestic sphere, to the high number of independent immigrant women, their organisations and contacts with groups such as the Catholic Church and trade unions. Nevertheless, being placed on the highest level of the political agenda does not of itself guarantee an appreciation of the range of immigrant women's aspirations and needs, as is demonstrated in the French case (Kofman 1997). The key programmes of the state agency dealing with the arrival and integration of female immigrants increasingly single out the early stages of life in France, leaving the needs of the more established immigrant women, such as for skills training to enhance employment opportunities, to mainstream programmes. Academically, the concerns and experiences of immigrant women do not figure, with some notable exceptions, on the agendas of mainstream feminisms.

CONCLUSION

As I have shown in this chapter, European states do not see the rights of migrants as human rights. Migration more than any other policy field belongs to the cherished remit of state sovereignty and states reserve the right to differentiate between migrants and citizens in areas such as employment and welfare. Coordination between European states has primarily aimed at the control of undocumented immigrants and an attempt to pass the buck further down the line to non-European Union states. The obsession with and manipulation of the numbers immigrating have meant that all flows that cannot be closely controlled have been subjected to increasingly restrictive regulation, as with family reunion and asylum. Thus whilst continuing to subscribe to human rights in a general way, the substantive content of these rights for migrants is being whittled away, and reduced to the minimum conditions, as is demonstrated by the resolution on family reunification proposed by the European Inter-Governmental Conference in June 1993.

At the same time, states are more than ever discriminating between the legal and the undocumented, the deserving and the undeserving, the established and recent arrivals. As pressure grows to reduce economic and social rights for nationals in European welfare states, so it is becoming more difficult to argue for recognising the needs and entitlements of migrants (Baubock 1993), although some initiatives for established migrants may be forthcoming from the European Union. It is unlikely, however, that an easing of flows will be envisaged, while legislation for denying even fundamental rights to the undocumented and, in effect, their children, has been enacted. Family reunification strikes at the very core of sovereignty and identity. It is not just about rights, but who can enter and on what conditions immigrant communities may expand in the future. Even the European Convention on Human Rights is veering towards the right of the state to pursue immigration control rather than ensure the right to family life.

It has not been my intention to present an unremittingly gloomy view of recent developments. The defeat in 1997 of right-wing governments in Britain and France, obsessed with immigration control, has brought some changes, such as the repeal of the primary purpose rule in Britain. Although the latter applied equally to men and women, it was always more difficult for women to bring in men, whose entry as potential workers was viewed suspiciously. Yet there is little sign of any loosening of the grip of dependency which casts its shadow over female family reunion migrants in particular. Given the tenacity of state sovereignty and migration control, we need to fight for more progressive policies in each state.

Family reunification is not just the Cinderella of human rights but of academic studies (Kofman 1999; Lahav 1996). I would argue that one of the main reasons for this neglect is that it has been associated with women as dependants following their male 'primary' migrant. The increasing masculinisation of family reunification and formation, and women as agents as opposed to dependants in this process, has passed almost unnoticed. Academic studies in themselves will not change policies, but we do need to understand the mechanism of these flows in order to suggest policies that reflect the diverse experiences and specific difficulties confronted by women in these kinds of migratory movements. The realisation that men too are imported as dependants may eventually help to decouple family reunion and female dependency. Above all, we have to work for policies which will give immigrant women greater independence. The first priority has to be to challenge the operation of probationary periods imposed by states which lock women in private patriarchies. Giving women access to skills training and legal employment will also enable them to lead more independent lives. The latter are some of the most pressing demands expressed by immigrant women's organisa-tions in Europe (European Women's Lobby 1995). After all, changing places internationally should not have to be equated with an extreme loss of rights.

NOTES

1 Currently it generally includes the spouse and children up to the age of 18 years (except for Germany in which it is up to 16 years). Most countries do not allow grandparents or recognise common-law (allowed in the UK until 1995) or same sex unions (except for the Netherlands).

REFERENCES

Ansay, T. (1991) 'The new UN Convention in the light of German and Turkish experi-ence', *International Migration Review* 25: 831–47.

Anthias, F. (1993) 'Gendered ethnicities in the British labour market', in H. Rudolph and M. Morokvasic (eds) *Bridging States and Markets*, Berlin: Sigma.

Anthias, F. and Yuval-Davis, N. (1993) *Racialized Boundaries: Race, Nation, Gender, Colour and Class and the Anti-racist Struggle*, London: Routledge.

Baubock, R. (1991) *Immigration and the Boundaries of Citizenship*, Monographs in Ethnic Relations 4, Warwick: Centre for Research in Ethnic Relations.

Baubock, R. (1993) 'Entitlement and regulation: immigration control in welfare states', in H. Rudolph and M. Morokvasic (eds) *Bridging States and Markets*, Berlin: Sigma.

Bhabha, J. and Shutter, S. (1994) *Women's Movement: Women Under Immigration, Nationality and Refugee Law*, Chester: Trentham Books.

Brah, A. (1993) 'Re-framing Europe: en-gendered racisms, ethnicities and nationalisms in contemporary Western Europe', *Feminist Review* 45: 9–29.

Campani, G. (1993) 'Labour markets and family networks: Filipino women in Italy', in H. Rudolph and M. Morokvasic (eds) *Bridging States and Markets*, Berlin: Sigma.

Campani, G. (1997) 'Women and social exclusion: the case of migrant women', paper given at the conference: Inclusion and Exclusion: International Migrants and Refugees in Europe and North America, organised by International Sociological Association, June, New York.

Crawley, H. (1997) *Women as Asylum Seekers: A Legal Handbook*, London: Immigration Law Practitioners' Association and Refugee Action.

Escrivá, A. (1997) 'Control, composition and character of new migration to South-west Europe: the case of Peruvian women in Barcelona', *New Community* 23(1): 43–58.

Essed, P. (1996) *Diversity: Gender, Color and Culture*, Amherst, MA: University of Massachusetts Press.

European Women's Lobby (1995) *Confronting the Fortress: Black and Migrant Women in the European Union*, Brussels: European Parliament.

Gaspard, F. (1994) *Obstacles in Society to Equality of Opportunity for Immigrant Women: The Situation in France, Belgium, Italy and Spain*, Strasbourg: Joint Specialist Group on Migration, Cultural Diversity and Equality of Women and Men, Council of Europe.

Ghai, Y. (1997) 'Migrant workers, markets and the law', in W. Gungwu (ed.) *Global History and Migrations*, Oxford: Westview.

GISTI (Groupe d'Information et de Soutien des Travailleurs Immigrés) (1994a) 'Famille, au ban de l'Europe', *Plein Droit* 25.

GISTI (1994b) *Entrée et Séjour des Étrangers: la nouvelle Loi Pasqua*, vol. 1, Paris: GISTI.

Goodwin-Gill, G. (1989) 'International law and human rights: trends concerning international migrants and refugees', *International Migration Review*, 23: 526–46.

Harzing, A.-W. (1995) 'The labour-market position of women from ethnic minorities: a comparison of four European countries', in A. Van Doorne-Huiskes (ed.) *Women and the European Labour Markets*, London: Paul Chapman.

Home Office (1996) *Control of Immigration Statistics United Kingdom 1995*, London: HMSO.

Hoskyns, C. (1996a) 'The European Union and the women within: an overview of women's rights policies', in R. Amy Elman (ed.) *Sexual Politics and the European Union*, Oxford: Bergahn.

Hoskyns, C. (1996b) *Integrating Gender: Women, Law and Politics in the European Union*, London: Verso.

Hoskyns, C. and Orsini-Jones, M. (1995) 'Immigrant women in Italy: perspectives from Brussels and Bologna', *The European Journal of Women's Studies*, 2: 61–76.

Hu Khoa, L. and Barou, J. (1996) 'Connaissances et usages du dispositif d'acceuil par les familles regroupées', *Migrations Etudes*, 68.

Hune, S. (1991) 'Migrant women in the context of the international convention on the protection of the rights of all migrant workers and members of their families', *International Migration Review* 25(4): 800–17.

Joint Council for the Welfare of Immigrants (JCWI) (1994a) *The Right to Family Life for Immigrants in Europe*, London: JCWI.

Joint Council for the Welfare of Immigrants (1994b) 'Family reunion policies in six European countries', *Eurobriefing* 1.

Joint Council for the Welfare of Immigrants (1996) 'European human rights disaster', *Bulletin*, p. 9.

Kofman, E. (1997) 'In search of the missing female subject: comments on French immigration research', in M. Cross and S. Perry (eds) *Population and Social Policy in France*, London: Cassell.

Kofman E. (forthcoming) 'Female birds of passage a decade later: gender and immigration in the European Union', *International Migration Review* 33.

Kofman, E. and Sales, R. (1992) 'Towards Fortress Europe?', *Women's Studies International Forum* 15(1): 29–39.

Kofman, E. and Sales, R. (1997) 'Gender and family reunification in the European Union: some considerations for refugee entry', *Refuge* 16(4): 26–30.

Lahav, G. (1996) 'International vs. national constraints in family reunification migration policy: a regional view from Europe', paper given at the International Studies Association Conference, San Diego, April.

Lebon, A. (1994) *Situation de l'immigration et présence étrangère en France 1993–1994*, Paris: La Documentation Française.

Leidholdt, D. (1996) 'Sexual trafficking of women in Europe: a human rights crisis for the European Union', in R. Amy Elman (ed.) *Sexual Politics and the European Union*, Oxford: Bergahn.

Lloyd C. (1998) 'Rendez-vous manqués: feminisms and antiracisms in France: a critique', *Modern and Contemporary France* 6(1): 61–73.

Lutz, H. (1994) *Obstacles to Equal Opportunities in Society for Immigrant Women, with Particular Reference to the Netherlands, the United Kingdom, Germany and the Nordic countries*, Strasbourg: Joint Specialist Group on Migration, Cultural Diversity and Equality of Women and Men, Council of Europe.

Morawska, E. and Spohn, W. (1997) 'Moving Europeans in the globalizing world: contemporary migrations in a historical-comparative perspective (1955–1994 v. 1870–1914)', in W. Gungwu (ed.) *Global History and Migrations*, Oxford: Westview.

Morokvasic, M. (1993) '"In and out" of the labour market: immigrant and minority women in the labour market', *New Community* 19(3): 459–84.

Morokvasic, M. and de Tinguy, A. (1993) 'Beyond East and West: a new migratory space', in H. Rudolph and M. Morokvasic (eds) *Bridging States and Markets*, Berlin: Sigma.

OECD (1993) *Trends in International Migration*, Paris: OECD.

Plender, R. (1991) *International Migration Law*, Dordrecht: Matinus Nijhoff.

Seifert, W. (1996) 'Occupational and social integration of immigrant groups in Germany', *New Community* 22(3): 417–36.

Southall Black Sisters (1997) 'The one year immigration rule: a stark choice: domestic violence or deportation?', *National Women's Network Newsletter*, July/August, p. 1.

Tribalat, M. (1996) 'Chronique de l'immigration', *Population* 1: 141–94.

Tribalat, M. (1997) 'Chronique de l'immigration: les populations d'origine étrangère en France métropolitaine', *Population* 1: 163–220.

United Nations (1991) 'International Convention on the Protection of the Rights of All Migrant Workers and Members of their Families', *International Migration Review* 25(4): 873–900.

Weber, F. (1993) 'European conventions on immigration and asylum', in T. Bunyan (ed.) *Statewatching the New Europe*, London: Statewatch and Unison.

9

DOES CULTURAL SURVIVAL HAVE A GENDER?

Indigenous women and human rights in Australia

Deborah Bird Rose

It may well be true that the history of women's equal rights, once it is written, will prove that retrogression replaced progress through a large part of what we consider to be the continuous advance of civilisation.

(Tomasevski 1993b: xii)

INTRODUCTION

The Indigenous people of Australia are an encapsulated minority in a settler society, a 'third world in the first' (Young 1995).[1] During two hundred years of colonisation, Anglo-Australian settlers have, through their racism and sexism, produced a situation of massive disadvantage for Indigenous people, particularly women. As I write, a new government has cut funding for programs for Indigenous people, is planning amendments to national land rights legislation (the Native Title Act), speaks nostalgically of the assimilation era, has refused both compensation and apology for former policies of enforced assimilation, and has massively undermined the reconciliation process that lies at the heart of human development for all Australians.

Human rights activism for the Indigenous peoples of Australia is concerned with two main aims: equal rights to health, education and employment, and at the same time, the right to survive as culturally distinct peoples. Thus, both equality and cultural difference are goals, giving rise to the potential for contradiction (Howard 1992; McDonald 1992). While government policy in recent decades has focused on the first set of concerns and resisted the second, many Indigenous people assert that there really is only one concern. They contend that as their societies, cultures and spiritualities are land-based, so human rights for them cannot be understood separately from land rights. Australia is committed to compliance with the International Covenant on Civil and Political Rights, Article 27 of which provides that groups must not be denied the right to enjoy their own culture. Aboriginal and Torres Strait Islander Social Justice Commissioner Michael Dodson (1995: 13) contends that cultural survival is inextricable from

140

land rights: 'Aboriginal cultures depend on our access to our homelands and our capacity to control what happens on our land.'

In principle, state and national governments support the concept of Australia as a multicultural society based on tolerance and formal equality, but the rights of Indigenous people remain particularly contentious. There is strong political pressure to ensure that Indigenous people's rights, particularly rights to land and resources, do not interfere with the objectives of non-indigenous people. From this perspective, cultural difference is tolerated for as long as it does not impinge on the lives of non-indigenous people.

My purposes in this paper are to examine the evidence with respect to the human rights of Indigenous people in Australia; to describe an Indigenous system of gendered power and knowledge as a basis for of cultural survival; to analyse some of the land rights contexts in which Indigenous women have been marginalised; and to reflect on the effect that the promotion of human rights has on cultural survival and gender social equity. I take the starting point identified so succinctly by Coomaraswamy (1994: 45), that human rights are about empowerment. The right to their land is identified by Indigenous people as central to their cultural survival. The argument is that colonising society imposes a set of gender constructs that marginalises and disempowers Indigenous women, even as it decolonises and reverses the original expropriation of land. The very practices which are formulated to empower Indigenous people have the potential actually to disempower women. This analysis links cultural survival and gender, and thus contributes to the feminist project which continues to expose the divergence between human rights formulated about a generalised 'people' and the specific human rights of women (Charlesworth 1995).

COLONISATION

The Australian nation came into being through blood, theft and pain. The first settlers arrived in 1788 to found a penal colony for the British empire, and Australia was for decades ruled by the musket and the lash. War was never declared and treaties were never made, but during the century of invasion and settlement (1788 to approximately 1900) Indigenous people suffered violent and widespread devastation. Across the continent the reduction of the Aboriginal population was about 90–95 per cent. Massacres, diseases, malnutrition, starvation, dispersal and assimilation all took a terrible toll as Aboriginal people were forced from their territories and their bases of subsistence, and either killed or concentrated in reserves, missions, cattle properties or fringe camps. Only in 1992, with the High Court's 'Mabo' decision, did Indigenous systems of land tenure become formally acknowledged by the conquering nation. Indigenous land rights are a crucial dimension of human rights currently being contested in Australia.

The history of conquest must inform any analysis of the current status and prospects of Indigenous people in Australia. Not only have they become a dispossessed minority, they have also been the target of official and unofficial policy and practice aimed at their elimination: in an earlier period through death and dispossession, and subsequently through enforced assimilation. The inquiry into the 'stolen generations' shows widespread and massive damage to families and individuals through the forcible removal from their families of children of mixed ancestry (National Inquiry 1997). Until 1967 Aboriginal people were not counted in the national census. They were not citizens, but rather wards of the state, and were denied fundamental human rights. In the early 1970s the federal government adopted an official policy of self-determination for Indigenous people, and thus only in the recent past has government policy sought to promote Aboriginal people's development without demanding assimilation. It remains the case that Indigenous people are regularly classed by others as people who constitute a 'problem' and for whom special legislation and policies are required.

Aboriginal people live in every type of Australian environment (urban, suburban, rural and remote; deserts, plains, coasts and islands). The human rights problems people face vary from place to place, but disadvantage remains significant throughout. Provision of special funding has been consistent with the well-accepted dictum that 'equal treatment of persons in unequal situations perpetuates rather than challenges discrimination' (Tomasevski 1993b: xiii). In 1996 a new government brought to the fore the kind of free trade thinking that holds that to assist any group more than another is discriminatory, and that all should be treated equally. This approach, applied to members of social groups who are patently not equal, is likely further to widen the gap between different social groups (Tomasevski 1993a: 12).

The achievement of land rights is a major step in reversing the history of dispossession. Several states and territories have enacted land rights legislation; as well, there is national legislation in the form of the Native Title Act (1993).[2] Additionally, the Aboriginal and Torres Strait Islander Heritage Act (1984) protects particular sites, and states and territories also have specific heritage legislation. As feminist analysis of human rights would have predicted, the laws were framed by men, the legal system is dominated by men, the anthropological discipline (brought into the field as an 'authority' on cross-cultural affairs and on Aboriginal people) was dominated by men, and land rights in Australia have marginalised Indigenous women in many cases. I will return to this issue shortly.

HUMAN RIGHTS IN QUANTITATIVE PERSPECTIVE

In June 1994 the population of Australia was about 17,843,300 people. Aboriginal people comprised about 1.7 per cent of that number. Population densities range from a low in Victoria, where Aboriginal people comprise 0.4 per cent of the population, to a high in the Northern Territory of 27.4 per cent (McLennan 1996b: 77, 113). Although numerically minuscule, the Indigenous presence looms large in the media, arts, politics, scholarship and popular culture.

The United Nations Development Program (UNDP) Human Development Report draws up a human development index based on life expectancy, schooling, employment and wage rates (Tomasevski 1993b: 56). I follow these criteria to make comparisons between Aboriginal men and women, and between Aboriginal and non-Aboriginal people.[3] The results show that the differences between Aboriginal women and men are nowhere near as pronounced as the differences between Aboriginal and non-Aboriginal people. A gross comparison between the two populations is unavailable, but the Australian Bureau of Statistics (1991–2 census) compares regions along a socio-economic index that measures relative disadvantage by combining figures concerning income, educational attainment and unemployment. A low score indicates that the area has more families of low income and more people with little training and in unskilled occupations than average. The lowest figure (726) is recorded for an area where the Aboriginal population is high, Bathurst and Melville Islands, Northern Territory (90 per cent Aboriginal). In contrast, the highest figure (1,110) is recorded for a well-to-do suburb of Canberra, the national capital, where Aboriginal people comprise 0.49 per cent of the population (Jain 1994: 56). A comparison of standardised mortality ratios by region shows precisely the same contour: high figures in areas of high Aboriginal population, and low figures in areas of low Aboriginal population (ibid: 57–60).

In 1991 the overall unemployment rate for Aboriginal people was 38 per cent, compared with a national figure of about 10 per cent. Among Aboriginal people, women were unemployed at about the same rate as men (Madden 1994: 45). Since 67 per cent of Aboriginal people who are in the workforce are now employed in the public sector, their dependence on government funding is considerable (ibid.: 46). The cuts, in August 1996, to government funding for Aboriginal communities and programs will undoubtedly drive up the rate of unemployment.

Aboriginal people are also disadvantaged in income. Considering Aboriginal people over the age of 15 in 1994, the average annual income was $A14,046: $A15,448 for men, and $A12,702 for women (Madden 1994: 48). This average was 30 per cent below that of the total population, which had an average annual income of $A20,000 (McLennan 1996a: 125). In addition, low incomes combine detrimentally with the high cost of living in remote communities where many Aboriginal people reside (Crough and Christopherson 1993: 40).

Universal education for Aboriginal people has only been in place for a few decades. In 1994, 7 per cent of Aboriginal people over the age of 15 had obtained a year 12 certificate, and 48 per cent had either no formal education or education below the year 10 level. In contrast, 74.6 per cent of non-Aboriginal Australians had completed year 12 (McLennan 1996b: 118, 279).

The poor health of Aboriginal people has already been foreshadowed. While their infant mortality rates have fallen since 1967 as a result of concerted efforts by government-funded initiatives, they remain much higher than the total Australian figures. In the Northern Territory, Aboriginal babies accounted for 73 per cent of infant deaths, but only 33 per cent of all births in the year 1993 (Bhatia and Anderson 1995: 12). It is predicted that on current trends, in the year 2011 life expectancy will be 62.3 years for Aboriginal males and 69.9 for Aboriginal females compared with 77 and 82 for non-Aboriginal males and females (ibid: 36).

Alcohol abuse and other forms of substance abuse have an impact on Aboriginal people at levels that appear to exceed that of the nation as a whole. Alcohol accounts for 27 per cent of the deaths of Aboriginal women, and although women consume less than men, in some areas more women than men die of alcohol-related causes (Brady 1991: 205). Male violence, often following alcohol consumption, is regularly suffered by women and children (Bolger 1991; Brady 1991: 204–5).

Rape of Aboriginal women is underreported and difficult to estimate for the same reasons as it is for women world-wide (see for example Kuzmanovic 1995). Racism and sexism combine in some areas to produce relatively unchecked sexual violence directed against Aboriginal women. Colonising myths of Aboriginal women's sexual availability (see McGrath 1984) conceal attitudes that would be better labelled genocidal rape (Copelan 1995).[4] A woman from Cape York (Queensland) stated: 'If a white woman gets bashed or raped here, the police do something. When it's us they just laugh. The fellows keeps walking around, everybody knows but nothing is done' (quoted in Paxman and Corbett 1994: 4).

Aboriginal men also commit rape, and it is sometimes the case that Aboriginal women are reluctant to report them, for reasons of kinship, community solidarity, and in the knowledge of the high number of male deaths in custody. While Aboriginal women are vastly overrepresented in prison, comprising 50 per cent of the female prison population in 1989 (Paxman and Corbett 1994: 4), deaths in custody are concentrated among men. In 1987 the federal government appointed a Royal Commission into Aboriginal Deaths in Custody. The Commission investigated 99 deaths, of which 11 were of women. The Commission concluded that the high proportion of Aboriginal inmates gave rise to the high incidence of Aboriginal deaths in custody. Overpolicing and oversentencing of Aboriginal people were identified as chief causes of incarceration (ibid.: 2).

In sum, the evidence indicates an overall picture of severe disadvantage. Statistics reliably cover only the last decade of two centuries of devastation, and although there are demonstrable areas of improvement, some recent trends suggest that in comparison with changes in the non-Aboriginal population, Aboriginal people are actually losing ground. So disparate are conditions for the majority of Aboriginal people and those of the majority of non-Aboriginal people, that they might as well be living in different worlds.

HUMAN RIGHTS AND SELF-DETERMINATION

Statistics tell a story of how Aboriginal people, aggregated into sets of figures, are differentiated from the Australian nation as a whole on the basis of what they do not have, and how they do not live. I turn now to an examination of human rights in the context of cultural survival. Here my focus is on what Aboriginal women do have and how they do live. This is a description of presence, and it presents

Aboriginal women in a very different light. I am not suggesting that one picture is more true than another, but rather that neither tells the whole story.

According to Michael Dodson (1996: 7), 'The most fundamental group right is the right to self determination. . . . Like all other peoples, we have the right to freely determine our political status and freely pursue our economic, social and cultural development.' This means that Aboriginal people have the right not only to attain equity with non-Aboriginal populations in quantitative terms, but also to develop their future in accordance with their own cultural values. The link between land tenure, human rights, gender and development is well understood internationally, as are the particular needs of Aboriginal and other peoples for whom land is not a commodity (Shiva 1993; Tomasevski 1993b: 35–7). In Australia, land is not only the basis of subsistence, but also (and at this time perhaps equally importantly), the basis of Aboriginal social and political autonomy, and spiritual well-being.

Aboriginal cultures across Australia (with the possible exception of some urban people who have been severed from their homelands for generations) construct identity, social relations and spirituality in relation to local place. Termed 'country', local place is intimately associated with the people who own it and belong to it within Aboriginal law. In the words of the Chairman of the Central Land Council, Rex Granites (1996: 1), 'Land rights are at the centre of Aboriginal concerns. Our understanding of country encompasses the full range of spiritual, cultural, economic, social and legal boundaries of our lives.'

It is clear that there were and are differences in gender relations in different parts of Indigenous Australia and under different colonising regimes. Through-out, however, male-dominated colonising society positioned women at the per-iphery and kept them there (Bell 1980: 264–5). There is a feedback effect such that women's marginal position under colonisation is often construed to be repre-sentative of the pre-colonial past. Androcentric accounts of Aboriginal societies which defined women primarily as objects of male control are now generally discredited, but the debate about Indigenous gender relations continues (see for example Bell 1980; Bell 1993; Bern 1979; Berndt 1950; Gale 1983; Hamilton 1981; Langton 1985; Merlan 1988; Merlan 1991; Rose 1992; Tonkinson 1990). Until recently Indigenous women have been largely excluded from the debates, and as they enter the arena their primary message, articulated forcefully by women such as author and activist Jackie Huggins, is that much of the debate has been racist and patronising (Riddett 1996).

My research experience has been in areas of the Northern Territory, Western Australia and New South Wales, where separation of men's and women's sacred domains ensures sites of power and autonomy such that men cannot dominate women and vice versa (Rose 1994). Like a number of other woman scholars, and like many Aboriginal women, I analyse gender relationships in a dialogical framework rather than one of domination. In examining power relations I look to those contexts in which women control their own lives, exercise their own spiritu-ality, and make decisions about themselves, their families, their country and their spiritual development. Gender-separated domains constitute the non-negotiable ground from which women are able to interact dialogically with men.

In Aboriginal Australia, the living world is a created world. As far as is known, throughout the continent cultural geography includes places that are gendered and sacred (see Brock 1989). Some places are managed jointly, but there are places to which men may go but of which they do not know the full meaning, and other places where men can never go. A similar landscape exists for men. The creating ancestors are often called 'Dreamings' in Australian English. Dreaming women and men imprint themselves on the earth, and leave behind the traces of their activities, the sites of their actions, and their specific presence. The created world does not privilege women to the exclusion of men, nor does it set women in opposition to men, although it does acknowledge the competitive quality of desire, including the desire to dominate. Rather, and far more profoundly, the creative beings, women and men, establish sites of autonomous power. The women who belong to and own the country have non-negotiable responsibilities toward the creative Dreamings whose actions, songs, ceremonies, and esoteric knowledge made and make this world. Country holds the gendered power which women unfold in their lives in dialogue with men.

In some parts of Australia the inheritance of knowledge has been severely curtailed as a consequence of previous decades of government policy, which aimed at assimilating Aboriginal people and thus eliminating their Aboriginality. In such areas, Aboriginal women are no less devoted than they were to the protection of sacred places, and to the control of knowledge which they transmit according to their own law. The fact that conquest has diminished the amount of knowledge they have been able to inherit from their ancestors enhances the significance of that which remains.

Indigenous cultural survival depends on country, knowledge and gendered responsibilities. That is, it is not just land that is at issue, but specific 'countries' with their own internal relations. Men and women have a joint interest in protecting those domains, as Aboriginal leader Mary Yarmirr (1997: 81) states: 'We are fighting beside men for recognition and for our country. Women still carry out the roles we have always carried out in relation to the cultural, physical, spiritual, environmental and social maintenance of our land, community and law.'

THE ERASURE OF WOMEN

Every place has a story, and virtually every place is a story; some of these stories and related knowledge are owned and known only by women, and some are only for men; many have multiple faces: a non-gendered face and one or more restricted faces. Land rights, heritage protection, and Native Title cases bring Anglo-Australian law together with Aboriginal law, creating the potential for harm as well as for good. On the one hand, Aboriginal systems of law establish separate gendered domains of place, knowledge and action; on the other hand, the Australian legal system asserts gender equity but is male-dominated in practice. Aboriginal law controls access to knowledge; Anglo-Australian law requires an open inquiry. The basis of formal equality is that opposing parties be treated equally, and that the proceedings be open so that fairness can be ensured.

In the context of land, the government requires an open inquiry into Indigenous ownership, and into the story or meaning of sites. Indigenous men and women are entrapped differently in this encounter. Male dominance in Anglo-Australian legal and political institutions generates situations in which non-indigenous men require women to discuss their knowledge in order to attempt to protect sites. In the case of the Alice Springs dam site, for example, the Northern Territory government planned to build a dam in an area that the Indigenous owners (Arrernte women) asserted must not be destroyed. The Minister for Aboriginal Affairs, Mr Tickner, appointed Mr Wootten, a distinguished member of the Australian judiciary, to investigate the matter. The Aboriginal women were placed in a double-bind: how to discuss the importance of the place with decision-makers who were men. Members of the Ngaanyatjarra Pitjantjatjara Yankuntjatjara Women's Council explained: 'It amounts to being forced to break our Law to prove to Europeans that our Law still exists . . . it threatens our culture, not just one or two individuals' (quoted in Wootten 1992: 74). This was one of several cases in which, in spite of strong opposition, there was a favourable outcome; the site is being protected (Wootten 1993).

A more recent and more highly contested case concerns development plans for a small bridge in South Australia, and opposition by Indigenous women who say that the bridge would destroy a sacred site, much of the knowledge of which is secret. This case went several times to appeal to the High Court of Australia; for several years women seeking to protect the site have endured massive public scrutiny of the legitimacy of their claim. A number of non-indigenous men, primarily journalists and academics (anthropologists and historians), and politicians, as well as some non-indigenous women and some Indigenous people, have asserted that this particular group did not have a tradition of secret women's law, and that therefore the current claim for the sanctity of the site could not be founded in tradition. Much of this debate centres on how one reads the ethnographic record for this area, and so it is possible, even probable, that the words of settler Australians will override those of Indigenous women. In this colonising context, race and gender combine, as Rao (1995: 174) argues for the international sphere, to produce 'a falsely rigid, ahistorical, selectively chosen set of self-justificatory texts and practices whose patent partiality raises the questions of exactly whose interests are being served and who comes out on top'.

In contrast to other states and territories, the Northern Territory has had land rights legislation since the late 1970s. The record of claims made under the Aboriginal Land Rights (Northern Territory) Act 1976 provides insight into the process whereby women are marginalised or erased. It is central to my argument that this legislation has been extremely beneficial to Aboriginal people (women included). In these years large portions of the Northern Territory (about 40 per cent of the land and 85 per cent of the coast, to date) have come under Aboriginal

Freehold Title, and large numbers of Aboriginal men and women have been found to be 'traditional owners' within the terms of the Act.

The Act requires that claimants give evidence to a specially appointed commissioner who is a federal court judge. In an astonishing number of claims, lawyers and anthropologists have deemed it to be quite adequate for men to speak for women and for women to say virtually nothing on their own behalf. In many other claims, women have voluntarily curtailed their evidence because many of the proofs of ownership were either restricted to women only, or were not suitable for scrutiny by men (see Bell 1984–5; Merlan 1991; Rose 1996a; Rose 1996b). Thus in many claims the evidence deeply obscures Aboriginal women's status and actions as managers of country, kinship and other social relations, as well as of their ecological, geographical, religious and other forms of knowledge. The spiritual dimension of their lives sometimes is not even mentioned.

The evidence reflects the male dominance of the legal profession (barristers and Land Commissioners have to date all been male) and the greater numbers of men who have been employed as senior anthropologists in the preparation of land claims, as well as a colonising history that assumes Indigenous male supremacy. It tends to reinforce the stereotype, commonly held by many non-Aboriginal men and women, that Aboriginal societies are male-dominated and that women are essentially pawns in social life. It also stands as testimony to a tunnel vision approach which asserts that in a world of tight budgets it does not matter who gives evidence as long as people get their land. In this view, gender equity appears to be classed as an optional extra that is simply not affordable. And as long as male centrality is reproduced because gender equity is implicitly defined as an optional extra, male centrality will appear to be the norm.

The erasure of the power and presence of women in a public hearing about the spiritual, cultural and social bases of land ownership is itself a form of violence; it obscures and tends to nullify the living presence of Aboriginal women in their social, moral and spiritual complexity. It is a violence that is not acknowledged as such because it is displaced; male supremacist views of society assume that women already are marginal to social and spiritual life. Given that assumption, their marginal position in land claims can be comfortably held to be a mirror of indigenous gender relations, rather than being seen for what it is, an instance of the violent suppression of these relations. This intensely serious and strongly legitimated setting establishes an enduring written record that can be understood to document a people's relationships to land at the time of the inquiry. The record may be read as a public benchmark of female absence; without intervention, the impression of absence is likely to increase rather than decrease.

CULTURAL SURVIVAL AND HUMAN RIGHTS

The fear of cultural extinction joins the fear of physical extinction as a critical issue to which people publicly dedicate themselves (Rudolph 1997: 5). The concept of human rights, while focused on the individual, includes collective or communal rights of cultural survival (Howard 1992; McDonald 1992; Svensson 1992). The core issue in Australia is whether Indigenous peoples will be able to continue to function within the contemporary Australian polity of which they are a part, and thus whether individuals will be able to enjoy their basic human rights.

Conflicts in which non-indigenous plans for economic advantage are challenged by Indigenous people seeking to protect sacred sites or to negotiate for the future of their lands, highlight these issues. The argument from developers suggests that tradition is a fossilised and outmoded world-view. Development ideology draws on the discourse of rights to contend that Aboriginality is a condition to be transcended. The logic is that tradition and human development are inescapably incompatible, and thus that human rights are best realised through assimilation rather than through adherence to customs and traditions. For Indigenous women there is a double entrapment here. To the extent that colonising institutions erase the living presence of women and promote facsimiles of their own patriarchy, women are disadvantaged. And to the extent that women are disadvantaged, colonising society can then claim a right to intervene to rescue women from the disempowering effects of male dominance. In this double entrapment, the greatest loss is women's personal and collective rights to assert the legitimacy of, and to control their own, institutions of power and knowledge.

Today most Aboriginal people assert their right to social equity in the major quality of life domains. At the same time, they also assert their human rights as Indigenous people to their cultural integrity, including at least some control over at least some of their lands. To the extent that Aboriginal women's presence and power is erased by Anglo-Australian institutions, it becomes more difficult for them to assert an autonomous power; and to the extent that assertions of autonomy fail, women are increasingly erased. Legislation designed to promote human rights for Indigenous people is put into practice by individuals who are themselves gendered, raced, and historicised subjects. As Bell (1992) states in an earlier discussion of some of these issues, law is applied neither in a vacuum nor in a mode of reflexive critique. To the extent that legislation directed toward cultural survival fails to include Indigenous women, it is difficult to see that long-term cultural survival will be achieved for men or women. And when gender equity is classed as an optional extra, all women are marginalised. Progress and retrogression go hand in hand, and the ideals of equality and respect for cultural difference are set on a far horizon.

NOTES

1 Throughout this paper I use the terms 'Indigenous' and 'Aboriginal' interchangeably to refer to Aboriginal and Torres Strait Islander peoples of Australia.
2 The government in 1998 is under pressure to amend the Act in ways that will decrease the impact it may have on non-Aboriginal people and corporations.
3 Figures are unreliable prior to the mid-1980s (to the extent that they exist at all), and even in the 1990s there are large numbers of non-stated or inadequately stated responses to census reports (Crough and Christopherson 1993: 19–21).

4 Relatively unchecked sexual violence is a legacy both of wars of invasion and of policies of assimilation; the latter is known by some as a policy to 'fuck em white', a term first documented around the turn of the century.

REFERENCES

Bell, D. (1980) 'Desert politics: choices in the "marriage market"', in M. Etienne and E. Leacock (eds) *Women and Colonization*, New York: Praeger.

—— (1984–5) 'Aboriginal women and land: learning from the Northern Territory experience', *Anthropological Forum*, 5(3): 353–63.

—— (1992) 'Considering gender: are human rights for women, too? An Australian case' in A. An-Na'im (ed.) *Human Rights in Cross-Cultural Perspectives, a Quest for Consensus*, Philadelphia, PA: University of Pennsylvania Press.

—— (1993) *Daughters of the Dreaming* (2nd edn), Sydney: Allen and Unwin.

Bern, J. (1979) 'Ideology and domination: toward a reconstruction of Australian aboriginal social formation', *Oceania* 50(2): 118–32.

Berndt, C. (1950) *Women's Changing Ceremonies in Northern Australia*, Paris: Hermann et Cie.

Bhatia, K. and Anderson, P. (1995) *An Overview of Aboriginal and Torres Strait Islander Health: Present Status and Future Trends*, an Information Paper, Canberra: Australian Institute of Health and Welfare.

Bolger, A. (1991) *Aboriginal Women and Violence*, Darwin: North Australia Research Unit.

Brady, M. (1991) 'Drug and alcohol use among Aboriginal people', in J. Reid and P. Trompf (eds) *The Health of Aboriginal Australia*, Sydney: Harcourt Brace Jovanovich.

Brock, P. (ed.) (1989) *Women, Rites and Sites*, Sydney: Allen and Unwin.

Charlesworth, H. (1995) 'Human rights as men's rights', in J. Peters and A. Wolper (eds) *Women's Rights, Human Rights: International Feminist Perspectives*, London and New York: Routledge.

Coomaraswamy, R. (1994) 'To bellow like a cow: women, ethnicity, and the discourse of rights', in R. Cook (ed.) *Human Rights of Women: National and International Perspectives*, Philadelphia, PA: University of Pennsylvania Press.

Copelan, R. (1995) 'Gendered war crimes: reconceptualizing rape in time of war', in J. Peters and A. Wolper (eds) *Women's Rights, Human Rights: International Feminist Perspectives*, London and New York: Routledge.

Crough, G. and Christopherson, C. (1993) *Aboriginal People in the Economy of the Kimberley Region*, Darwin: North Australia Research Unit.

Dodson, M. (1995) Native Title Report, July 1994–June 1995 (Report of the Aboriginal and Torres Strait Islander Social Justice Commissioner), Canberra: Australian Government Printing Services.

—— (1996) 'Human rights for Aboriginal peoples', *Northern Analyst*, 1: 6–8 (published by North Australia Research Unit).

Gale, F. (ed.) (1983) *We Are Bosses Our-selves*, Canberra: Australian Institute of Aboriginal Studies.

Granites, R. (1996) 'Foreword', in programme for the conference 'Land Rights, Past Present and Future', held 16–17 August, Canberra.

Hamilton, A. (1981) 'A complex strategical situation: gender and power in Aboriginal Australia', in N. Grieve and P. Grimshaw (eds) *Australian Women: Feminist Perspectives*, Melbourne: Oxford University Press.

Howard, R. (1992) 'Dignity, community, and human rights', in A. An-Na'im (ed.) *Human Rights in Cross-Cultural Perspectives, a Quest for Consensus*, Philadelphia, PA: University of Pennsylvania Press.

Jain, S. (1994) *Trends in Mortality*, catalogue no. 3313, Canberra: Australian Bureau of Statistics and National Centre for Epidemiology and Population Health.

Kuzmanovic, J. (1995) 'Legacies of invisibility: past silence, present violence against women in the former Yugoslavia', in J. Peters and A. Wolper (eds) *Women's Rights, Human Rights: International Feminist Perspectives*, London and New York: Routledge.

Langton, M. (1985) 'Looking at Aboriginal women and power: fundamental misunderstandings in the literature and new insights', paper presented at ANZAAS Conference, Monash University, Melbourne, August 1985.

Madden, R. (1994) *National Aboriginal and Torres Strait Islander Survey*, Catalogue no. 4190, Canberra: Australian Bureau of Statistics.

McDonald, M. (1992) 'Should communities have rights? Reflections on liberal individualism', in A. An-Na'im (ed.) *Human Rights in Cross-Cultural Perspectives, a Quest for Consensus*, Philadelphia, PA: University of Pennsylvania Press.

McGrath, A. (1984) ' "Black Velvet": Aboriginal women and their relations with white men in the Northern Territory, 1910–1940', in K. Daniels (ed.) *So Much Hard Work: Women and Prostitution in Australian History*, Sydney: Fontana/Collins.

McLennan, W. (1996a) *Australian Social Trends*, ABS catalogue no. 4102, Canberra: Australian Bureau of Statistics.

—— (1996b) *Year Book of Australia 1996*, ABS catalogue no. 1301.0, Canberra: Australian Bureau of Statistics.

Merlan, F. (1988) 'Gender in Aboriginal social life: a review', in R. Berndt and R. Tonkinson (eds) *Social Anthropology and Australian Aboriginal Studies, a Contemporary Overview*, Canberra: Australian Institute of Aboriginal Studies.

—— (1991) 'Male–female separation and forms of society in Aboriginal Australia', *Cultural Anthropology* 7(2): 169–93.

National Inquiry into the Separation of Aboriginal and Torres Strait Islander Children from their Families (1997) 'Bringing them home', Sydney: Human Rights and Equal Opportunity Commission.

Paxman, M. and Corbett, H. (1994) 'Listen to us: Aboriginal women and the white law', *Criminology Australia* 5(3): 2–6.

Rao, A. (1995) 'The politics of gender and culture in international human rights discourse', in J. Peters and A. Wolper (eds) *Womean's Rights, Human Rights: International Feminist Perspectives*, New York and London: Routledge.

Riddett, L. (1996) ' "Finish, I can't talk now": Australian Aboriginal and settler women construct each other', Department of History, University of Saskatchewan, Occasional Paper No. 3.

Rose, D. (1992) *Dingo Makes Us Human; Life and Land in an Australian Aboriginal Culture*, Cambridge: Cambridge University Press.

—— (1994) 'Flesh and blood, and deep colonising', in M. Joy and P. Magee (eds) *Claiming Our Rites: Studies in Religion by Australian Women Scholars*, Wollstonecraft, NSW: Australian Association for the Study of Religions.

—— (1996a) 'Histories and rituals: land claims in the Territory', in B. Attwood (ed.) *In the Age of Mabo: History, Aborigines and Australia*, Sydney: Allen and Unwin.

—— (1996b) 'Land rights and deep colonising: the erasure of women', *Aboriginal Law Bulletin* 3(85): 6–13.

Rudolph, S. (1997) 'Introduction: religion, states, and transnational civil society', in S. Rudolph and J. Piscatory (eds) *Transnational Religion and Fading States*, Boulder, CO: Westview Press.

Shiva, V. (1993) 'The impoverishment of the environment: women and children last', in M. Mies and V. Shiva (eds) *Ecofeminism*, Melbourne: Spinifex.

Svensson, T. (1992) 'Right to self-determination: a basic human right concerning cultural survival. The case of the Sami and the Scandinavian state', in A. An-Na'im (ed.) *Human Rights in Cross-Cultural Perspectives, a Quest for Consensus*, Philadelphia, PA: University of Pennsylvania Press.

Tomasevski, K. (1993a) *Development Aid and Human Rights Revisited*, London: Frances Pinter Publishers.

—— (1993b) *Women and Human Rights*, London: Zed Books.

Tonkinson, R. (1990) 'The changing status of Aboriginal women: "free agents" at Jigalong', in R. Tonkinson and M. Howard (eds) *Going it Alone: Prospects for Aboriginal Autonomy*, Canberra: Aboriginal Studies Press.

Wootten, H. (1992) 'Significant Aboriginal sites in the area of proposed Junction Waterhole Dam, Alice Springs', Report to the Minister for Aboriginal Affairs.

—— (1993) 'The Alice Springs dam and sacred sites', *Australian Quarterly* 65(4): 8–22.

Yarmirr, M. (1997) 'Women and land rights: past, present and future', in G. Yunupingu (ed.) *Our Land is Our Life*, St Lucia: Queensland University Press.

Young, E. (1995) *Third World in the First, Development and Aboriginal Peoples*, London: Routledge.

10

WOMEN AND HUMAN RIGHTS IN POST-COMMUNIST COUNTRIES

The situation in the Czech Republic

Jiřina Šiklová *

Several months after the coup of November 1989 and restoration of the democratic system, the then federal state of Czechoslovakia was visited by representatives of feminist and women's activist groups from other countries who wanted to inform us about human rights. At that time, after forty years of almost total isolation from the West, we listened, receptive to all that anyone could teach. One of the women lecturers kept saying 'human rights are women's rights', and she advised us to teach other women articles of women's rights, maybe by having them sing the chorus 'human rights are my property' while working in the rice fields. We pointed out to her that we had neither rice fields nor cotton plantations, that Czechoslovakia was situated in Central Europe and had long been among the advanced industrial nations, that only 4 per cent of the population worked in agriculture, that it was 100 per cent literate, and that women, on average, had higher levels of education than men.[1] The lecturer then changed her remarks, but only marginally.

It dawned on me at that time that we are not, from the feminist point of view, just younger and poorer 'stepsisters' of our counterparts in the West, but that we have our own experience with the emancipation of women and feminism, which is so far ideologically unreflected and unspecified. Certainly we women who have experienced and survived socialism have something to say to the West (Šiklová 1995a, 1996)! In both blocs, the view of the other has been distorted, owing to isolation, sealed borders and ideological struggles. On both sides there is a lack of understanding and shared memory. One of these communication gaps concerns 'human rights' and another the idea of feminism and human rights as 'women's rights'. In this chapter, I will first explore the ways in which 'human rights' were conceptualized under the socialist regime and comment on reasons why women have not taken up the call to associate 'women's rights' with 'human rights' in the

* The author wishes to thank Janice Monk for her assistance in editing this chapter.

post-socialist Czech Republic. I will then explore two areas in more depth, women's rights in employment and women's rights in relation to their bodies, identifying issues that activists and planners need to address if women's human rights are to be better realized within a post-Communist state.

HUMAN RIGHTS AND WOMEN'S RIGHTS UNDER THE SOCIALIST REGIME

The emancipation of women under the socialist regime was made 'from the top'. It was conceived in limited terms, and identified as the duty of women to work in some profession in addition to taking care of the children and the home. To use the terminology of the time, it was women's duty to 'participate in building socialism'. Feminism in this period was either not mentioned at all or was officially denounced by Marxist-Leninist philosophy as a bourgeois ideology intended to confuse the working class, to split it into two groups pursuing particular interests, thereby postponing the victory of the proletariat and the building of socialism (Butorova 1996; Cermakova *et al.* 1995, Wolchik 1991).

Our approach to human rights was contradictory in a similar way. We did not know them, were not allowed to discuss them or to cultivate and refine our attitudes. Simultaneously, the term 'human rights' was 'occupied' and appropriated by the ruling political regime. In the 1960s when human rights and civil rights were being most strongly fought for in the West, those words that ostensibly have the same sense in East and West in fact took on completely different meanings.[2] For nearly half a century it was proclaimed in the socialist bloc that enforcement of individual human rights was superfluous, as a way of demonstrating the victory of the proletariat: a classless society was supposed to have realized the human rights of all poor people and of all marginal groups, including women and ethnic minorities. Women's rights, minority rights and other social issues were considered temporary problems that would be automatically solved by the establishment of socialism. The interests of an entirely abstract collective society or social system were clearly put above the importance of the individual, while the subject of history was usually represented as people, all working people, or the proletariat, always disregarding differences of sex, nationality or race.

For the most part, human rights, like feminism or the right of nations for self-determination, were simply not discussed. At times they were interpreted as the 'trap of capitalism' and as pseudo-democracy. Later, in 1975, during the process of approximation or convergence between the West and the East, Czechoslovakia, together with the other countries of 'real socialism', entered the Helsinki Accord on Civil and Political Rights and on Economic, Social and Cultural Rights. These agreements came into force in Czechoslovakia in March 1976, but they were not observed, either in this country or in the other socialist countries of the time. Indeed, texts of the agreements were not published, even though they had been officially adopted.[3] No photostating was available, dissidents had one much-handled copy circulating amongst them, and people were persecuted if they made copies and attempted to circulate them.[4] Under these circumstances, the broad public could learn nothing about the declarations of human rights or discuss them openly. Yet the suppression had the effect of generating opposition by dissidents; resistance movements came into being with the purpose of forcing their governments to respect the human rights enunciated in the agreements.[5] The concept of human rights became a symbol of resistance in socialist countries, although almost no one knew exactly what they were at the time.

In the 1980s, when Gorbachev was in power in the Soviet Union, governments of socialist countries became more interested in cooperation with the West and their representatives sometimes spoke about human rights, especially at meetings with foreign officials. At that time, whoever it was who was speaking about human rights was more important than the content of the statement. That is, it was more important whether the declarations were referred to by someone from the socialist establishment of the time or by someone from the dissident groups. The schizophrenia went so far that, for example, in the summer of 1988 the Czechoslovakian government itself tried to establish the League for Human Rights, while interfering with all the meetings of the internationally established Helsinki Committee led by the dissident Jiři Hájek.

In this period, there was no difference between men and women in their attitudes toward human rights. Those who were interested in them, men and women alike, called for and appealed to them. After the coup in 1989, the texts were published several times, but they were not really studied, so that even now we do not know exactly what human rights are, although they are discussed a lot and *believed* to be well known.

The attitude towards the slogan 'human rights are women's rights' is similarly schizophrenic. Women's issues (feminism and the emancipation of women) and human rights are two terms that were ideologically contaminated for decades, contradictorily interpreted and used both by the establishment in socialist countries and by dissidents. To change our thinking is a major task. It will be a long process to organize our thoughts, get rid of prejudices and misinterpretations, to learn about human rights and various alternative feminisms, to learn tolerance, and to free ourselves from the totalitarian visions that are so deeply rooted. We have to accomplish these tasks ourselves as citizens, as women. They cannot be taught by anyone from 'outside'; putting a slogan about human rights into the chorus of a song is not a solution for us.

The issue of women's rights was ignored not only by the socialist establishment but also by the opposition. Neither Charter 77, as the oldest, most clearly formulated dissident protest, nor Solidarity in Poland, Fides in Hungary, and the Movement for the Support of Human Rights in the Soviet Union (begun by, Sakharov) paid attention to woman's rights. The dissidents emphasized primarily those parts of international declarations on human rights that concerned freedom of speech, the right to disseminate information, the right to plurality of political parties and opinions, and the abolition of censorship, but they ignored the position of women and their human rights. They considered them to be 'particular'

155

problems which did not have to be settled at the time of bondage, when 'more important issues of freedom of all people' were at stake. Essentially they thought within the same scheme as the representatives of the existing regime: when democracy prevails, the problem of the position of women and their human rights will be solved as well.

The example of the oldest and most distinct group of opponents of the political regime in the former Czechoslovakia, Charter 77, illustrates the common pattern. Despite being constantly banned during the thirteen years of its existence, Charter 77 produced 573 documents about various social and political problems. These documents pointed out serious problems of the time (Skilling 1981), criticizing censorship, the education system, low pensions and health care; they asserted the right to profess religious faith and for Romany minority rights, but not a single document was devoted to the issue of women in socialism. In none of today's post-Communist countries did the struggles against totalitarian regimes make any claims for women's rights; conclusions from discussions of general human rights were not applied to the position of women nor were women's issues even considered. Women themselves, although extensively represented among dissidents, did not promote these issues. They thought that the right time had not come. They behaved unwittingly like women in the 1960s on campuses in the USA, who participated in the fight for minority rights without realizing that they themselves were in the position of being a minority that was being discriminated against within a rebellious group. Because real political terror prevailed in the socialist countries at that time, disputes between dissidents would have been a luxury.

Why is it that women are not interested in their rights even at present, when human rights are so often discussed? Why must certain inadequacies and inequalities in the position of women in post-Communist countries be pointed out and described by experts from the West, while women in these countries fail to raise the issues? There are several explanations, not all of which can be included here. During the 'real socialism', women belonged to the 'working-class', to the pseudo-privileged group. Socialism was built in the name of (and allegedly in favour of) the working class; workers and children from working-class families were preferentially enrolled in schools to acquire a better social position; the state boasted about having a 'working-class president'; the term 'working-class government' was used frequently. Those in power then legitimized their rule as well as the terror (expressed, for example, by teaching about the dictatorship of the proletariat) by claiming that they were defending the interests of the 'oppressed of this country', including not only workers but women.

The reality was different. Both the working class and women were substantially worse off than those in the West in similarly advanced industrial societies, but in comparison with the situation before World War II, the position of workers, as well as of women, did improve dramatically. Despite the fact that women in the early 1950s were called to join in with the 'building of socialism', mostly because the economy needed cheap labour, this mass entrance of women into the labour force included the incorporation of women into professional fields and related increases in enrolments of women in colleges and universities. This change in the status of women, directed 'from the top', really influenced the position of women in the former Communist states and had an impact on women's self-evaluations (Scott 1975).[6] The regimes had quotas not only for workers but also for women in various schools and professions, as well as for representation in Communist Party bodies. In Czechoslovakia, women made up almost 30 per cent of delegates of all Congresses of the Communist Party and were similarly represented in parliament. They exerted no influence over political life, however. At that time, everything was decided by the political bureaus of the Party, or possibly by its Central Committee which was subject to the Politbureau of the Communist Party of the Soviet Union. The relatively large participation of women in public functions was therefore just a formality.

Many western feminists are unable to comprehend this system, and point out that women used to be better represented in the politics of socialist countries than they are at present. They fail to understand that though women were indeed in the parliament, decisions were not made there. The incorporation of women and workers into the existing political bodies on the basis of predetermined quotas resulted in these two social groups becoming discredited; this situation has had contemporary reverberations. Any suggestions of affirmative action for women are subconsciously associated with this phase of the so-called emancipation of women to such an extent that even the Social Democratic Party has denied that it is pursuing a pro-woman policy. In the elections in June 1996, women in the Czech Republic actually voted for the right-wing parties to a greater extent than men.[7]

The former practice of electing women into political bodies just to have women represented is probably the main reason why women in politics today publicly and repeatedly proclaim that they do not want to pursue 'a woman's policy' and represent women's interests, but rather focus on the programmes of the political parties to which they belong. An extensive analysis of newspaper articles and in-depth interviews with female politicians by a member of Gender Studies Prague, M. Vodrazka (1996), reveals that even female politicians themselves do not realize (and do not want to realize) and acknowledge what is or might be a specifically female approach to public matters.

During the coup (approximately from November 1989 until the elections in 1990) women were more strongly represented in the political movement and participated more extensively in forming democratic structures than today. Later they were placed on lists of candidates in practically non-electable positions and did not get the support of the boards of political parties that had nominated them. Unfortunately, even women aspiring to the role of politician did not protest against this procedure. On the contrary, they seem to accept their ousting from politics with a certain satisfaction. In the election period between 1992 and 1996, 10 per cent of members of parliament were female and there were no female ministers. After the elections of 1996, although the number of women in parliament increased, the government of the Czech Republic seems to have a 'male cabinet' again, having no woman member. In this respect, the situation in Slovakia is somewhat better, with a higher representation of women in parliament and the government (Butorova 1996).

Currently, citizens are beginning to realize that the number of women in politics should be higher; according to an opinion poll in 1996, 72 per cent of citizens would prefer to see more women elected. But no one, not even women themselves, is particularly sensitive to the clearly discriminatory attitudes that signal that the status of women continues to be inferior. For example, before the elections to the Senate (in October 1996) the invitation leaflet addressing the electorate was phrased in the masculine gender, implying that only males were being asked to vote. The mistake provoked no protest by women nor a discussion or apology on the part of the organizers. As recently as 1996 the report of the Czech Helsinki Committee included a chapter covering some aspects of women's rights in the Czech Republic which stated that 'over the past two years the interest in the Czech public in the issue of women in politics has increased', but we are far from being at the stage at which women were in western countries when they started to enter politics in the early 1970s.

Although the absence of women in supreme political bodies is becoming recognized, the consciousness of the general public and of women politicians needs to be raised to incorporate a bigger range of issues. Because 'feminism' and 'the women's movement' have such negative connotations in the Czech cultural context, ways need to be found to present issues in other terms (Czech Helsinki Watch 1996). This is a difficult task since we lack obvious issues of wide appeal. Certain rights such as the right to abortion, maternity leave, employment, and broad access to education have been attained, albeit imperfectly, under socialism, while other, more subtle general issues that might unify women have not yet been recognized.

Newly established women's organizations tend to focus on concerns specific to particular subgroups of women, such as the problems of single mothers, domestic violence, the problems of mothers of disabled children, or the issue of leisure time of children whose mothers go out to work. Some groups are based in religious denominations and completely avoid politics (*Aspect* 1997; Busheikin 1994; Havelkova 1995). A further problem is that women's movements and organizations in the Czech Republic, Slovakia and other post-Communist countries do not have broad membership and are ignored by the majority of women. Individual women's organizations do not support each other, they do not even know about each other and do not associate and coordinate their actions. Any lobbying by women's organizations is so far unthinkable in post-Communist countries (Gabal 1995; Havelkova 1995; Šiklová 1996).

Under the current conditions in the Czech Republic it is very difficult to secure grants for studying the position of women. The Institute for Public Opinion Polls has only recently begun to take notice of the topic. It is true that gender studies have been taught at four universities in the Czech and Slovak Republics, but only thanks to the initiative of the non-governmental organization, Gender Studies Prague, which has the support of funds from a German foundation (Bollag 1996; Šiklová 1995). The Ministry of Education, university rectors and the administrations of colleges and universities continue to consider teaching of this discipline superfluous, although they do not admit that in front of partner organizations in the West. The university boards do include women, but even they – probably because of the fear of being labelled feminists – do not promote this teaching. To put it briefly, some of the enemies of women's issues are the women who behave as a minority that does not want to be seen too much (Dahlerup 1985; Matynia 1994). Such psychological barriers in women cannot be eliminated by a legal decree or affirmative action but only by systematic, ordinary, totally unrevolutionary work and almost daily reminders of the problems.

If women's human rights are to be advanced in the Czech Republic, two fields for action can be clearly identified: efforts to change the conditions for women in employment, and the development of new perspectives on women's rights in relation to their bodies. In both areas, new concepts need to be articulated and new mechanisms developed for addressing problems.

WOMEN'S RIGHTS IN RELATION TO EMPLOYMENT

Three broad issues need to be addressed for women to have equal rights in relation to employment: access to jobs and promotion; equal pay with men; and problems of sexual harassment.

The Czech Republic in 1996 had almost full employment, with an unemployment rate of only 3.5 per cent and little difference between the unemployment rates of women and men. A review of labour legislation conducted in 1996 in order to assess the prospects of including the Czech Republic within the European Union (Castle-Kanerova 1996) found that equal rights to work and remuneration are in harmony with the Declaration of Fundamental Human Rights ratified by Czechoslovakia in January 1991, and with the Convention to End All Forms of Discrimination of the International Labour Organization. At the same time, it notes that in view of the recent history of the totalitarian regime in the country, affirmative action, that is, positive discrimination in favour of promoting women into decision-making positions, is unacceptable to the general population and to women themselves.

Despite the *de jure* absence of discrimation, however, *de facto* discrimination exists in the workplace. Employers have no obligation to declare that they provide 'equal opportunity' and few employers or employees understand what the concept might mean. Neither women nor trade unions have ever protested against public advertising requiring only men to apply to a position of manager, while secretaries must be 'young, attractive women'. Rather, such facts are pointed out by foreign women journalists or visitors. Nobody documents and monitors cases of indirect discrimination against women and proposals by Gender Studies Prague that such monitoring should be carried out by a non-governmental organization, independent of the government, are totally disregarded.

The culture of employment reflects an ideology that men are 'breadwinners' who cannot settle for a salary that is 'appropriate for a woman', and further looks at women workers in relation to their home and family caring roles. Employers often refuse to hire women with small children for more exacting jobs. Women applying for positions are asked about the ages of their children or whether they plan to have children. The situation has its roots in the former regime. While women were expected, and expect themselves, to be in the workforce, employers automatically assumed that women would perform at a lower level, since they had to take care of children and family, and therefore would not aspire to higher positions. Even in professions where women were in the majority, for example, in pediatrics, where women represented 76 per cent of all graduates, only 18 per cent held executive positions in 1986 and they did not hold decision-making positions. These attitudes towards women workers have a positive effect in that they create a non-competitive atmosphere between women and men in the work-place, but a negative effect in that women are considered to be less effective workers and their responsibility primarily as familial caretakers is legitimized (Castle-Kanerova 1996).

Many women, in order to be able to combine work and family roles, accept jobs as home workers. As such they enter relationships not covered by collective agreements, which include about 40 per cent of employees. Rather, they hold individual work contracts which do not provide legal protection. Such circum-stances are not defined as problems by employers, trade unions, or women workers.

The segregated labour force that has resulted leads to a differentiation in male and female salaries (Cermakova *et al.* 1995) and lower prestige for female-dominated jobs, such as nursing, teaching, medical doctors, and other jobs in which women form the majority. When salaries, wages and even extra bonuses are determined, distinctions are made on the basis of sex, with the higher salary or bonus automatically awarded to men. Women are seen as secondary wage earners and accept this view. Survey research conducted by the Sociological Institute of the Academy of Sciences under the direction of Marie Cermakova (1996) and a poll by the Institute of Public Opinion show that both women and men are aware of gender inequality. For example, 51.6 per cent of women believed that men have better career opportunities and men also recognize this fact. The research reports that 65 per cent of women and 64 per cent of men are convinced that women have less opportunity than men to become director of a firm or an executive. Similarly, both sexes are aware to the same extent that women have less opportunity than men to secure high-paying jobs (42 per cent of women and 41.4 per cent of men). Of respondents who think they have experienced on-the-job discrimination, only 2.5 per cent of men think they were discriminated against because of sex, whereas 39.2 per cent of women report experiencing sex discrimination. Despite the difference in awareness of discrimination, there seems to be a kind of labour 'reconciliation': sexual inequality in most work situations was not rated as a major problem and neither study reported attempts to push women out of the job market and into the home or direct cases of discrimination. Such survey results, although they appear in print from time to time, provoke almost no debate (Cermakova 1996; Wolchik 1994). In my opinion, these relationships echo the attitudes created in the past when men and women were 'allies' against a paternalistic, totalitarian state hostile towards its citizens (Funk and Mueller 1994).

After the coup in 1989, many prognosticators in the West expected women to withdraw quickly to the home and to cease to aspire to higher education. This has not happened yet. The number of women at universities, colleges, and in other post-secondary institutions has not decreased and girls have not been discriminated against from the point of view of school enrolment.[8] For the future, however, we have substantial reason to fear that when the educational system is privatized to a greater extent than at present, families will prefer to invest in the education of their sons rather than daughters, given acceptance of expectations of greater returns.

The laws of the post-Communist countries also lack significant provisions against sexual harassment in the workplace. An offender can only be prosecuted when the sexually harassed woman can prove that, owing to her resistance to his sexual advances she was subject to some disadvantage at work, for example, that she was denied a bonus or reduced to a lower rank, or when the offence was a criminal infringement on personal freedom or oppression. Blackmail can also be included within this realm of offences. Verbal abuse, ridicule, sexually tinged jokes or proposals do not fall within legal jurisdiction. Most women would not report such behaviours, however, since they are not aware that they are an attack on their human rights.

WOMEN'S RIGHTS IN RELATION TO THEIR BODIES

Women's rights to health are ensured by general health insurance, which is compulsory for all citizens and continues to function in the post-Communist state. In this respect, women's rights are identical to men's. These conditions also apply to pensions schemes, that is, financial and health security in old age. Women have the right to retire five years earlier than men in the event that they have given birth to at least two children. Though women in post-Communist countries live, on average, nine years longer than men, because of the automatic valorization of pensions they are relatively secure financially in old age.

Reproductive rights and freedom of behaviour in sexual matters are also mostly ensured in these countries. State approaches to abortion changed several times as socialism developed. At the beginning of the socialist era, abortion was prohibited; it existed only illegally and was severely punished. Later, legislation underwent substantial change. It was possible to obtain an abortion with the approval of a special commission consisting of laymen which was also often a form of political supervision. At the end of the socialist period, non-surgical termination of pregnancy within eight weeks was allowed and was used almost as a form of contraception. All abortions were performed in state health facilities and were covered by health insurance. Since the coup in 1989, attitudes towards abortion have not changed markedly. Only the woman makes the decision either to keep the baby or to have an abortion, and the opinion of the father does not have to be respected. Today abortion must be paid for (the cost represents half the average monthly salary) and the woman has to take vacation time from work to have it performed; her absence cannot be covered by pay. As a result of these changes in policy, the number of abortions in the Czech Republic has dropped, and a similar trend can be observed in Slovakia. Abortion policies are similar in all the post-socialist states, with the exception of Poland, although the quality of service differs.

In general, human rights in relation to other aspects of sexuality, as well as reproductive rights, are not repressed in the Czech Republic; it has standards close to those of western countries. The situation reflects the relatively low religiosity of the country and a generally liberal approach to sexuality.[9] Lesbian rights are not restricted, and lesbians and gays are not limited in their activities or discriminated against in public. Prejudice against them is relatively low-key. Public debates on the topic are quite frequent and public acceptance is greater than is acceptance of feminism. Both lesbians and gay men are now pursuing the possibility of entering into legal matrimony, of having inheritance rights, and rights to adopt children. Lesbian and gay journals are published in the Czech Republic and prosper quite well (Renna 1997).

Newly established non-governmental organizations, helped by foreign women's organizations and foundations (especially East-West Women's Network, Open Society/Soros, Frauen Anstiftung (Hamburg) and the Heinrich Böll Foundation (Germany)), have organized several conferences and public discussions about domestic violence, violence in the streets, rape, sexual trafficking (i.e. trade with women and the situation of prostitutes in border areas). Numerous studies on these themes and on the sexual abuse of children and teenagers have been conducted and reported in the media. Such problems were not discussed at all under socialism, so that in this respect, a taboo has been lifted.

The issues have aroused public as well as governmental interest from the Ministry of the Interior, the Ministry of Labour and Social Affairs, and other bodies. The problems have been reported and monitored and appeared for the first time in 1996 in the report of the Czech and Slovak Helsinki Committee. This report (the 22nd Report of the Helsinki Committee) states that the occurrence of rape and attempted rape has increased, and police statistics, which are only of reported cases, confirm this: 545 cases in 1989 and 726 cases in 1996. Police reports, however, are estimated to cover only approximately one-third of occurrences. Violence against women in the Czech Republic and Slovakia, and in other post-Communist countries, goes unreported. Women do not know their rights, are afraid to make reports to the police, do not know how to proceed, and have no idea what constitutes violence and sexual harassment. Particularly striking is the lack of knowledge and fear of persecution for the whole family in the Soviet Union, as testified in novels that can only now be published, as well as studies such as those by Alexander Solzhenitsyn, or collections on the position of women under socialism published in the West (Wolchik 1994).

In the Czech Republic, the issue of domestic rape only began to be discussed after the first public debate on television initiated by Gender Studies Prague and Profem, non-governmental organizations. From a stormy response to this debate, we learned that 90 per cent of women kept domestic rape by their husband, mate or lover a secret, and were not aware that it is a criminal offence under Article 241. They do not file a complaint because they have no idea that they can complain, and they do not know where to report such a case. Those who did report met with great lack of understanding on the part of investigators, and courts and police were not prepared to handle this type of investigation. This circumstance again reflects conditions of the past. The family was seen as an 'asylum' against the omnipresent 'Big Brother' of the socialist state establishment. As a result, rape is only reported by women with very strong and balanced personalities (Czech Helsinki Watch 1996).

The Czech Helsinki Committee, as well as Gender Studies Prague, considers it most important that the debate has been opened up and that a new attempt at verbalizing women's human rights has been initiated. In a pro-active vein, a number of advice centres have been established for women in crisis and on the run from home as well as an advice centre for refugees and for women who find asylum in the Czech Republic. Several relevant non-governmental organizations have also been created in the Czech Republic, Slovakia, Poland and Hungary.[10]

WORKING FOR CHANGE IN THE CONTEXT OF THE TRANSITION FROM THE SOCIALIST STATE

The 1996 report of the Czech Helsinki Committee correctly observed that women's rights must be put into effect taking into account Czech social and cultural traditions, and the existing degree of knowledge and perception of human rights and women's rights by women themselves. While serious discussions have begun within non-governmental organizations, the next stage requires attention at the governmental level.

Democratic states and their organizations have an obligation not to limit or interfere with the exercise of individual rights and also to find ways to uphold them. This requirement of the state is, however, relative. The relativity stems from the fact that the quality and quantity of what the state can provide depends on the economic prosperity of the state and the political system. The content and amount of legal claims will be determined by the usual law and by executive provisions in conformity with that law, which should not contradict principles of rights stipulated in international conventions. While regulations in force in post-Communist countries do not contradict the international conventions, they are not fulfilled as they should be. The situation reflects the struggle which these countries are having to find a democratic way of operating, to transform their economies from a command system to a privatized market system. In these circumstances, the state is being deprived of the important functions that it used to hold, and at the same time new demands are being made on it. Among those demands is that the state become the guarantor of the observance of human rights, including social (as well as political) rights. Meeting this requirement is very difficult for a state undergoing transformation. The state is required to become both weaker (especially in relation to the economy) and stronger (in relation to guaranteeing individual rights).

From this point of view, Czech and Slovak societies are relatively well off, because at least they can follow traditions of the period 1918–38 when Czechoslovakia was the only really democratic country in central Europe. In contrast, the majority of post-Communist countries lack any democratic period in their histories, so their economic and political transformation can only follow the pattern of other democracies.

All countries formerly belonging to the Soviet bloc leapt from feudal monarchies into the state called the 'dictatorship of the proletariat', and later declared themselves socialist states. During their transformation by means of violent coups, property transfers and changes in ownership of the means of production, nobody cultivated democratic thought or taught people (either men or women) to formulate their own requirements or organize themselves in a political or civic way. The *coups d'état* that took place at the end of 1989 are thus neither palace coups nor a mere taking over of power by another group, as has been common in revolutions in Latin America and some African countries. The transformation is different from those of totalitarian states where there was continuation of the capitalist system, both for society as a whole as well as for its constituent groups, including women.

In post-Communist countries, not only did the ruling dictatorial group change to a democratically elected one, but total restructuring of ownership was implemented. Nationalized property was re-privatized (in more or less suitable ways), and restitution made. Citizens in these post-Communist countries are learning not only to formulate their political programmes and elect democratic governments, but also to be economically independent, to organize themselves and be responsible for themselves. They are dealing with the abolition not only of totalitarianism but also of state paternalism. Those who are relatively strong, fierce, competitive and capable of private enterprise wish for a very weak state, a mere administrative body. Those who are relatively weak – and this usually includes women – wish for at least partial preservation of state paternalism in order to have the state guarantee their rights and social claims. Those who demand guarantees of human rights from the state do not wish to see the state weakened too much. They are consequently in conflict with those who demand a total limitation of state interference and the transfer of all powers to citizens. The socialist state denied human rights of all citizens, not only women by, for example, forbidding them to travel abroad, obtain and disseminate information, write and speak without censorship, establish associations, organizations and political parties. Now, after such a short period, the state is to become a guarantor of these human rights. The change is too sudden, not only for the state and its officials, but also for its male and female citizens. The barriers between the state and its citizens – in our case women – consist of lack of knowledge of civil and human rights, of civil obligations toward the state and the administration and the ignorance of group needs, requirements and existing possibilities on the part of women.

CONCLUSION

Women's lack of understanding of the concepts of human rights, and specifically of women's rights, remains a formidable barrier to change in post-Communist states. Women themselves have not clarified their requirements and positions because they see themselves as part of a society that is transforming itself as a whole. They do not know how to prioritize at this moment.

In Czech and Slovak society, 68 per cent of all citizens have changed their social status. They have changed their occupations, interests and roles because structures have been changed. Companies have been re-privatized, factories and fields have found their owners again and employees have become private entrepreneurs. Women in post-Communist countries were incorporated into society during socialism according to the status they acquired through their own efforts and that status has substantially changed. They are strongly influenced by these changes in what might be termed their 'secondary' status more obviously than in their 'primary' status as women and possibly mothers.

Under socialist regimes, all citizens, not only women, learned passivity and conformity. While women are now aware that women's rights, as part of human rights, must be incorporated into the existing social, political and national structures of the world, they also know from bitter experience that these structures can be abused and distorted in various ways. There is also an awareness that if post-Communist countries adopt the system of western democratic countries, they will take on the faults of those systems, which include limitations of women's rights.

Because of their previous experience (and survival) of coups and revolutions, women in post-Communist countries are too realistic to believe that, on the basis of their particular women's interests, they can create a new world order that eliminates existing unfair social arrangements and the inferior position of some groups. Therefore their attitude towards promotion of human rights is more restrained than that of women who lack this experience and who live in stable democratic states where changes are continually accomplished.

Women's relationship to human rights as women's rights is ambivalent because those of them who are interested in these issues know that the guarantor of human and social rights must be a strong state with a well-functioning administration and well-paid civil service. Such a state has just been dismantled. The totalitarian socialist state was on the one hand an enemy, while on the other it promoted employment, education and qualification of women, preferred women as a biological group, celebrated women as mothers of future soldiers and protectors of the fatherland and used them for its ideological and nationalist objectives. Turning to such a state as an upholder of one's claims seems paradoxical and difficult for women. Someone must proclaim and enforce women's rights, guarantee their legal and economic observance and protect those who promote women's rights against the rights of others. The protector of such a 'space' is unfortunately always the state or another supranational body. In view of their experience, women are afraid of the power of such a guarantor.

Women from post-Communist countries therefore find themselves at the beginning of a very long journey as they reflect on their new position and formulate their claims for human as well as women's rights. At the same time, they have to learn to address others who are at a disadvantage, gain their support and put forward their interests within the political system.

In addition to the above mentioned scepticism, women from post-Communist countries are full of prejudices against themselves, which is sometimes a danger for them. Educating women so that they will understand and seek fulfilment of human rights, including women's as fully as men's, is the fundamental challenge to be met if women are to step out of this vicious circle. It is up to us to find our own way, however, not simply to adopt or take over a prefabricated western model.

NOTES

1 The post-Communist states are not, and never were, a homogeneous bloc. Prior to World War I, what is now the Czech Republic, industrially well developed, was one of the ten richest states in Europe. It was relatively undamaged during World War II and maintained a high standard of living compared to the other socialist states under totalitarian regimes. The transition to privatization has also been less severe. In 1995, the Gross Domestic Product of the Czech Republic was the equivalent of $10,500 per capita compared with $7,300 in Slovakia, and about $6,000 in Poland and Hungary (Czech Republic Statistical Office 1995).

2 Despite the fact that Vienna is 250 km to the east of Prague, Austria belongs to the 'West' because it has remained a capitalist country. Because Czechoslovakia fell under Soviet influence in 1948 it has been taken to be part of 'Eastern' Europe (Musil 1995).

3 The Universal Declaration of Human Rights was reprinted in the Czech language by the UN in May 1969; the first edition was suppressed in 1970. It was next published in Prague in 1991.

4 Charter 77 Declaration, (1 January 1977). Charter 77 is in violation of Czechoslovak laws (see Skilling 1981: 212–13). On cooperation with the Human Rights Movement in Poland, see Skilling p. 277, and with Andrei Sakharow in the Soviet Union, see Skilling, p. 279. See also Šiklová (1996).

5 The Helsinki Citizens Assembly started in 1987. Helsinki Watch in Czechoslovakia was founded in 1988 and the League for Human Rights started in spring 1989.

6 The female labour force as a percentage of the total labour force for the years 1980, 1990, and 1994 was as follows:

Czech Republic: 47.2 per cent, 47.1 per cent, and 44.6 per cent;
Hungary: 43.4 per cent, 48.6 per cent, 47.4 per cent;
Poland: 45.4 per cent, 45.3 per cent, 46.3 per cent;
Slovak Republic: 45.5 per cent, 46.8 per cent, 46.2 per cent (Paukert 1995).

A special issue of *Transition: Open Media Research Institute: Events and Issues in the Former Soviet Union and East/central and Southeastern Europe* 1(16) 1995 also provides information about women in Russia, post-Soviet Ukraine, Hungary and the Czech Republic.

7 In the parliamentary elections of 1996, 6.6 per cent more women than men voted for right-wing oriented parties and 9.4 per cent fewer women than men voted for the Social Democrats (INFAS: Factum Praha 1996).

8 Education by sex in the Czech Republic, 1988 and 1994, by percentage:

Educational level achieved	1988			1994		
	Total	Men	Women	Total	Men	Women
Primary	22.6%	15.9%	31.2%	12.1%	8.5%	16.2%
Vocational/technical	41.0%	49.0%	30.6%	45.9%	53.0%	37.6%
Secondary	26.7%	23.3%	30.9%	31.9%	26.5%	38.1%
Tertiary	9.7%	11.8%	7.2%	10.1%	11.9%	8.0%

Sources: Microcensus 1989; Labour Force Survey, Prague 1994.

9 The Catholic Church estimated that only about 11 per cent of the population are believers, though 46 per cent are formally members of the Church (see Schuler 1995).

10 The Gender Studies Prague Library and Curriculum Center was founded in 1991. Profem (a non-governmental organization for the support of women's proposals) started in Prague in 1995; the White Circle of Security (in the Czech Republic) in 1994 (see *Altos and Sopranos* 1994).

REFERENCES

Altos and Sopranos: A Pocket Survey of Women's Organizations in the Czech Republic (1994), Prague.
Aspect: Quarterly for Feminist Culture (1993) Bratislava, no. 1.
Bollag, B. (1996) 'Women's studies programs in Eastern Europe', *The Chronicle of Higher Education* 12: 22.
Busheikin, L. (1994) 'I don't know what feminism is, but I say No!', in *Butterbrod and Bodies*, Prague: Centre for Gender Studies.
Butorova, Z. (ed.) (1996) *She and He in Slovakia: Gender Issues in Public Opinion*, Bratislava: Focus.
Castle-Kanerova, M. (1996) *Evaluation of Equal Possibilities for Men and Women in the Czech Republic*, Prague: Ministry of Work and Social Policy.
Cermakova, M. (1996) 'Women and men in the labor market – feelings of discrimination between men and women', *Data and Facts* 7: 1–4.
Cermakova, M. *et al.* (1995) 'Women, work and society', Working Paper 95, no. 4, Prague: Institute of Sociology.
Czech Helsinki Watch (1996) *Some Aspects of Women's Rights*, Prague: Czech Helsinki Watch.
Dahlerup, D. (1985) *Unfinished Democracy* Oxford: Pergamon Press.
Funk, N. and Mueller, M. (eds) (1994) *Gender, Politics and Post-Communist Reflections from Eastern Europe and the Former Soviet Union*, New York: Routledge.
Gabal, I. (ed.) (1995) *Ethnic Relations in the Czech Republic: the Czech Population's Attitude towards Foreigners*, Prague: New Technologies.

Havelkova, H. (ed.) (1995) *Gibt es ein mitteleuropäisches Ehe- und Familienmodell?*, Prague: Theatre Institute.

Malinova, H. (1993) *Development of Prostitution after 1989, Prostitution from the Point of View of Law in the Czech Republic*, Working Paper, no. 4.

Matynia, E. (1994) 'Women after Communism: a bitter freedom', *Social Research* 61(2) (Summer): 251–77.

Musil, J. (ed.) (1995) *The End of Czechoslovakia*, Budapest: Central European University Press.

Paukert, L. (1995) *Economic Transition and Women's Employment in Four Central European Countries 1989–1994*, Geneva: International Labour Office.

Renna, T. (1997) 'A lesbian: an interview with Jana Stepanova', in T. Renna (ed.) *Ana's Land, Sisterhood in Eastern Europe*, London: Westview Press.

Schuler, M. (1995) *From Basic Needs to Basic Rights: Women's Claim to Human Rights Women*, Washington DC: Law and Development International.

Scott, H. (1974) *Does Socialism Liberate Women?* Boston, MA: Beacon Press.

Šiklová, J. (1995) 'Political background: the Czech Republic', *Prelude, New Women's Initiative in Central and Eastern Europe and Turkey* 16–20.

Šiklová, J. (1996) 'Report on women in post-Communist Central Europe: personal view from Prague', Foreword to Z. Butorova (ed.) *She and He in Slovakia: Gender Issues in Public Opinon*, Bratislava: Focus.

Skilling, H. G. (1981) *Charter 77 and Human Rights in Czechoslovakia*, London: George Allen and Unwin.

Vodrazka, M. (1996) *Dialogue about Feminism*, Prague: Center for Gender Studies.

Wolchik, S. (1991) *Czechoslovakia in Transition, Politics, Economics and Society*, London: Frances Pinter Publishers.

Wolchik, S. (1994) 'Women and the politics of transition in the Czech and Slovak Republic', in M. Rucscherncyer (ed.) *Women in Post-Communist Eastern Europe*, London: Sharpe.

Part IV

CONCLUSION

11

GENDER, PLANNING AND HUMAN RIGHTS: PRACTICAL LESSONS

Tovi Fenster

Is it possible to draw conclusions that will contribute to the human rights discourse, after this journey around the globe through such different geographical, political, economic and cultural settings? I argue that along with the fundamental differences between places there are features in common. They give us practical examples and may guide us in deciding how better to implement human rights in planning and development in a way to benefit women.

GENDER AND POWER IN PLANNING AND DEVELOPMENT

Planning and development of the built environment is a very powerful process because of its influence on key aspects of everyday life, in which women's human rights, such as their access to resources, employment, housing, welfare and political participation, are frequently abused. One of the main arguments, which cuts across all chapters, is the significance of women's empowerment in planning and development in order to ensure that women's human rights are met. Access to power means access to knowledge and to involvement in the decision-making processes in planning and policy-making.

Jo Little explores this issue in discussing women's representation in planning bodies in the UK. Her analysis shows how the failure to meet women's needs in the built environment relates to lack of women's access to power, to decision-making and to knowledge about planning in some city councils in the UK. One important conclusion that can be drawn from this situation is that the form and shape of decision-making about the built environment must be looked at especially from the point of view of its patriarchal nature.

Gillian Davidson suggests that women's human rights can be met only if equality and rights come before, or together with, economic and material growth. If women take more control over their lives and are accorded more respect there is bound to be confrontation, both on a domestic and a public level. The state, instead of empowering women, has brought about a deterioration of their status.

171

It is a difficult project, but if women's rights are to be respected as human rights, the starting point must be recognition, knowledge and identification of violations against women and the opportunity to record them as such. It is only then, Davidson says, that inequality, difference and exclusion can begin to be addressed and eliminated, and women can begin to emerge from the margins of development and planning.

Another example, which highlights the importance of women's empowerment in planning and development, is the case of the Indigenous women in Australia. In this case, power is defined as the access to and influence on development plans, which conflict with the preservation of sacred sites of Indigenous people or do not take into consideration their land rights. The challenge here is to address the compatibility or incompatibility of tradition and human development. Indigenous women are doubly trapped. Colonizing institutions promote their own patriarchal norms and disadvantage women, and then claim a right to intervene to rescue them from the disempowering effects of male dominance. This situation negatively affects women's personal and collective rights to assert the legitimacy of, and to control their own, institutions of power and knowledge.

In searching for reasons why women in the post-socialist Czech Republic have not taken up the call to associate 'women's rights' with 'human rights' Jiřina Šiklová also brings in the notion of power. She points out that the number of women in politics is not high enough but stresses that no one, not even women themselves, is particularly sensitive to the clearly discriminatory attitudes that signal that the status of women continues to be inferior. Thus we see that empowerment is not only an external dynamic but mostly an internal process which can be carried out by the women themselves.

UNIVERSALIST-MODERNIST PLANNING AS A CONTROL ON WOMEN

The discourse around universalism versus relativism in relation to human rights has a strong impact on planning and development. One of the conclusions that can be drawn is that this dilemma must be tackled at the outset of planning. The impact of each approach on cultural minority groups and especially on women should be carefully evaluated prior to promoting plans for such groups. Tovi Fenster's analysis shows how taking a 'universalistic' view, which ignores cultural norms of a community, can bring about a deterioration in women's well-being even if these norms traditionally subordinate women. Her research reveals how closely power relations are connected to meeting women's human rights. In this case power, and especially male power, has an effect on women's freedom to move in space and to maintain cultural habits. Planning can become a tool for control: control of the state over minority groups and control of both state and men over women in minority groups. This patriarchal structure clearly jeopardizes the articulation of gendered human rights in planning. It emphasizes the necessity of including a dialogue among community members themselves and between them and planners in order to identify the needs and wants of the different members of the community, their cultural perceptions of space and the plurality of choices if social change is to occur smoothly while meeting their rights.

These notions of culture, human rights and planning are also raised in the case of the Indigenous women in Australia. Deborah Rose analyses the discourse about Indigenous people's land rights and women's land rights as a cultural concept that the colonizing society 'universalized' in the sense of imposing a set of gender constructs that marginalize and disempower Indigenous women. The two case studies she presents make clear how lack of consideration by colonizing planners of cultural and gendered constructions of Indigenous people negatively affected women's status and position.

PLANNING FOR DIVERSITY AS A MEANS OF MEETING HUMAN RIGHTS

Acknowledging the cultural, economic, gender, class and social diversity of cities in the modern world is not a common practice among planners. Such acknowledgement, however, can open the way to better understanding and implementation of human rights of different groups within cities and the built environment. Marcia Wallace and Beth Moore Milroy focus on this problematic situation of planning for difference and diversity and suggest two options. First, they propose incorporation of diversity into traditional planning practice on a case-by-case basis. This option improves access for minority groups to the planning process through measures such as increased outreach and accommodation of specific demands made by pressure groups. The second option they suggest is to take diversity as a point of departure. In this view, the city is seen as already constructed by various ethnic, class, gendered and age groups. It demands an understanding that diversity is fundamental to whoever we are already. This second approach is obviously more compatible with meeting human rights issues in planning and development.

WOMEN'S ECONOMIC RIGHTS AND DEVELOPMENT

Women's economic rights in development are closely connected to the notion of power, especially in the domestic spheres. Ann Oberhauser explores the links between women's economic status in Appalachia and human rights, arguing that unless women's status is improved, certain households, communities and regions will continue to lag behind. Development planners should build on analyses of household gender relations and divisions of labour to explore alternative means of income-generation, especially among women in low-income households. She mentions a few examples of community-based projects, which could promote women's economic rights; micro-enterprise development, flexible manufacturing networks and rotating loan programs.

Development policies such as structural adjustment programmes do affect the economic, social and cultural rights of women, as shown in Maureen Hays-Mitchell's analysis of these policies in Peru. The major conclusion she draws is that simply including women in programmes of micro-enterprise development will not be sufficient to overcome the discrimination they sustain from structural adjustment. It is necessary to reconsider the premiss on which structural adjustment operates in order to take into account non-economic factors that contribute to the gender-specific experience of structural adjustment in societies such as Peru. Thus, micro-enterprise development is but a preliminary step towards the more complete empowerment of women. But before final determination can be made, analysis of the long-term impact of, and problems associated with, all types of micro-enterprise development is needed. Moving women's rights into the centre of the human rights discourse in development has several components. Firstly, men's non-economic control of women should be analysed. Secondly, the changes should be identified in gender relations and social systems that are necessary to include the voices, experiences, insights and perspectives of women who have been excluded from the benefits of human rights and development. Thirdly, analyses should be made of the particular historical and cultural settings in which people in Latin America and other places in the world live and work.

GENDER AND MIGRANT RIGHTS

An important conclusion in Eleonore Kofman's chapter is that states actually do not consider migrant rights as human rights, and thereby reserve the right to differentiate between migrants and citizens in areas such as employment and welfare. Although some changes are being made in Britain and France as a result of the defeat of right-wing governments, still more progressive policies in each state must be implemented. One of the reasons why the revision of the family reunification laws is neglected is that these laws are associated with a view of women as dependants following their male 'primary' migrant. Regrettably, the increasing masculinization of family reunification and formation has almost passed unnoticed in academic studies. Academic studies will not change policies but they help to create an understanding of the mechanism of these flows. This is necessary in order that policies can be suggested that reflect the diverse experiences and specific difficulties confronted by women in these kind of migratory movements. Above all we have to work for policies which will give immigrant women greater independence.

The purpose of this book has been to explore the geographies and spatialities of human rights as they relate to gender, planning and development and to examine the significance of these components in various countries in order to identify ways to promote human rights. The authors have shown that women need power, access to knowledge and decision-making in the promotion of human rights in planning and development. Acknowledgement of cultural diversity and a better understanding of non-economic issues in promoting women's economic rights and advocating women's migrant rights, are equally crucial issues for planners. This book only begins to shed some light on these complicated problems, and it is hoped that it can provide assistance to all those who are devoted to the idea of human rights for all people in their diversity.

INDEX

Kaplan, S.: and Westheimer, R. 47
Knesset 40
Kong, L. 76
Kymlicka, W. 70–1n

Labour Party 36–7
land rights 140–1, 145, 147–8
language training programs 63, 71n
Latin America 99
Lee Kuan Yew 75
legal profession 148
lesbian rights 162
Lister, R. 6
local authorities: women and decision-
  making in 29–35, *30*; women's
  committees 16–17
local politics: and planning initiatives for
  women 35–7
local state: attack on 35–6

male bias 5
marginalization: of women 96
marriage 41, 127, 131; West Virginia 105
Marxist-Leninism 154
Mary's story 104–5
Massey, D. 79
menstruation hut 46–8, 51–2
methodology 38n
micro-enterprise development: Peru 112,
  113–16, 117–20, 120–1, 122n, 174
middle class 36
migrants: European Union 125–39,
  174–5; reunification 126–7, 128, 130–1,
  132–4, 174; rights 129–35;
  undocumented 126, 127, 129, 136
migration 19; East–West 128; *see also*
  immigration
minorities: national/ethnic *see* ethnic
  minorities
Mohanty, C. T. 96, 98
Moser, C. O. N. 11, 52, 86
multicultural societies 74
multiculturalism: Canada 56, 58–62
murder: of women 41
Muslim Bedouins *see* Bedouin women

Nairobi Forward-looking Strategies for the
  Advancement of Women (1985) 7
Nanavut 72n
National Advisory Council on the Family
  and the Aged (Singapore) 83
national minorities 70–1n

national values 18
needs: strategic gender 52
neo-Kantianism 11
New Urban Left 30, 35
non-governmental organizations (NGOs):
  Czech Republic 162; Peru 115, 116,
  122n
Northern Territory (Australia) 147–8
Northwest Territories 72n
Nussbaum, M. 49

Oberhauser, A. M.: Waugh, L. J. and
  Weiss, C. 105–6
occupations *see* employment
Ontario 63; Human Rights Commission
  63; Municipal Board 67; planning
  66–9; Planning Act (1996) 66–7
Ordinary Levels (O-Levels) 82, 86n

PAN (Practice Advice Note from RTPI)
  28–9
participation 5–6, 12
patriarchy 3, 7, 12, 41, 45, 172; Peru 112,
  114–15
People's Action Party (PAP) 74, 76–8, 81
persecution: gender-specific 129
Peru 111–24; structural adjustment
  programme 18–19, 111–12, 113–14,
  173–4
Peters, J.: and Wolper, A. 95
planning: classification 9–10; definition 8;
  Muslim Bedouin/Ethiopian Jews 42–8;
  profession 32
Poland 128
population planning 80–3
positive discrimination 68, 159
poverty: feminization 113–16; West
  Virginia **103**, 103
powers: Canadian land use planning 65
prison 144
private/public discourse 8, 12–14, 15, 79,
  95
pro-natalist programmes 82
procedural planning 9
progressive planning 9
Property Standards by-laws: Canada 64
public bodies 67
public spaces 3, 12; Bedouin 45, 51, 52
purdah 50
purity customs 46–8

Qadeer, M. 70